Lecture Notes in Mathematics

Edited by A. Dold and B. Eckmann

542

David A. Edwards
Harold M. Hastings

Čech and Steenrod Homotopy Theories with Applications to Geometric Topology

Springer-Verlag
Berlin · Heidelberg · New York 1976

Authors
David A. Edwards
Department of Mathematical Sciences
State University of New York at Binghamton
Binghamton, N.Y. 13901/USA

Harold M. Hastings
Department of Mathematics
Hofstra University
Hempstead, N.Y. 11550/USA

Library of Congress Cataloging in Publication Data

Edwards, David A 1946-
Čech and Steenrod homotopy theories with applications to geometric topology.

(Lecture notes in mathematics ; 542)
Bibliography: p.
Includes index.
1. Homotopy theory. 2. Geometry, Algebraic. 3. Algebra, Homological. 4. Algebraic topology. I. Hastings, Harold M., 1946- joint author. II. Title. III. Series: Lecture notes in mathematics (Berlin) ; 542.
QA3.L28 vol. 542 [QA612.7] 510'.8s [514'.24]
76-40180

AMS Subject Classifications (1970): 14F99, 14G13, 55B05, 55D99, 55J99

ISBN 3-540-07863-0 Springer-Verlag Berlin · Heidelberg · New York
ISBN 0-387-07863-0 Springer-Verlag New York · Heidelberg · Berlin

Printing and binding: Beltz Offsetdruck, Hemsbach/Bergstr.

to Marilyn and Gretchen

CONTENTS

§1. INTRODUCTION

Inverse systems of topological spaces occur in many contexts in topology. Some examples are:

a) Geometric Topology. One can associate to any embedding $X \hookrightarrow Y$ the inverse systems $\{U\}$, $\{U \setminus X\}$, $\{(Y \setminus X, U \setminus X)\}$, etc., where U varies over the open neighborhoods of X in Y. If X is locally compact, then one can associate to X its end, $\varepsilon(X) = \{X \setminus C\}$, where C varies over compact subsets of X.

b) Algebraic Topology. The Čech construction associates to a topological space an inverse system of CW-complexes. A similar construction in algebraic geometry leads to étale homotopy theory. The Postnikov and pro-finite completion constructions also associate inverse systems of complexes to complexes.

It is often important to have an appropriate category and homotopy category of inverse systems of spaces. In [Gro-1] Grothendieck showed how to associate to any category C another category pro-C whose objects are inverse systems in C indexed by "filtering categories" and whose morphisms are so defined as to make cofinal systems isomorphic. In [Q-1] Quillen introduced the notion of a model category as an axiomatization of homotopy theory on Top and SS. A model category is a category C together with three classes of morphisms called cofibrations, fibrations and weak equivalences which satisfy "the usual properties." The homotopy category of C, Ho(C), is obtained from C by formally inverting the weak equivalences in C. In [A-M] Artin and Mazur developed the algebraic topology of pro-Ho(C). One has a canonical functor pro-C $\longrightarrow$ pro-Ho(C) and it is natural to consider pro-Ho(C) as the homotopy category of pro-C. This point of view goes back to Christie [Chr]. But Christie also realized that for some purposes pro-Ho(C) was too weak and one really wanted a stronger category. It can be shown that pro-Ho(C) is not the homotopy category of a model category structure on pro-C. The second author [Has-1] has shown that pro-SS admits a natural model category structure with homotopy category, Ho(pro-SS), obtained from pro-SS by formally inverting level weak equivalences. Ho(pro-SS) had previously appeared in Porter's work on the stability problem for topological spaces [Por-2]. Grossman [Gros-1] has studied a coarser model category structure on Towers-SS.

In the first part of these notes (§§2-5) we develop the algebraic topology of Ho(pro-C) and compare Ho(pro-C) with pro-Ho(C). We also give applications to the study of the derived functors of the inverse limit. The second part of these notes (§§6-8) contains applications to proper homotopy theory, group actions on the Hilbert cube, and shape theory. We conclude with a brief list of open questions in §9.

More precisely, §2 contains background material about pro-categories and model categories. The "Mardesic trick," described in §2.1, says that all inverse systems are pro-equivalent to inverse systems indexed over cofinite strongly directed sets. Quillen's theory of model categories [Q-1] is reviewed in §2.3.

In §3 we show that for nice closed model categories C, C^J (where J is a cofinite strongly directed set) and pro-C inherit natural closed model structures from C, extending [Has-1].

In §4 we show that the natural inclusion Ho(C) $\longrightarrow$ Ho(pro-C) has an adjoint holim: Ho(pro-C) $\longrightarrow$ Ho(C) (compare Bousfield and Kan [B-K]) and obtain vanishing theorems for $\lim^s$.

The basic algebraic topology of Ho(pro-C) is developed in §5. We compare Ho(pro-C) with pro-Ho(C), discuss homotopy and homology pro-groups, and prove various Whitehead theorems. §5.5, Whitehead and stability theorems, includes a survey of work of the first author and R. Geoghegan [E-G-1-5].

In §6 we show that the category of σ-compact spaces and proper maps may be embedded in a suitable category of towers which is a closed model category. We then apply pro-homotopy theory to proper homotopy theory and weak-proper-homotopy theory (see [Chap-1] and [C-S] for weak-proper-homotopy theory and its uses in the study of Q-manifolds and shape theory.) Some of our results are announced in [E-H-3].

We apply this theory in §7 to the study of group actions on infinite dimensional manifolds. §7 represents joint work with Jim West [West-1], [E-H-W].

In §8 we discuss strong shape theory and develop generalized Steenrod homology theories using pro-homology. These theories have found recent applications in the Brown-Douglas-Fillmore theory of operator algebras [B-D-F-1-2]; see also Kaminker and Schochet [K-S], and with D. S. Kahn, [K-K-S].

Detailed introductions precede §§3-8.

Acknowledgements. We wish to acknowledge helpful discussions with Tom Chapman, Ross Geoghegan, and Jim West, and correspondence with A. K. (Peter) Bousfield and Jerry Grossman. Some of this material was presented at conferences at Syracuse University (Syracuse, N. Y., U.S.A., December, 1974 and April, 1975), Mobile, Alabama, U.S.A., (March, 1975), New York (March, 1975), the University of Georgia, U.S.A., (August, 1975), Guilford College (Greensboro, N. C., U.S.A., October, 1975), and the Interuniversity Center in Dubrovnik, Yugoslavia (January, 1976). We wish to acknowledge the organizers of these meetings for their hospitality.

The second-named author held a visiting position at the State University of New York at Binghamton during the academic year 1974-75 in which much of this work was done. He wishes to acknowledge their support and hospitality. He was also partially supported by N.S.F. Institutional Grants at Hofstra University in 1973-74 and 1975.

We wish to thank Althea Benjamin for typing this manuscript.

§2. BACKGROUND

§2.1. Pro - Categories.

We need a category of inverse systems such that cofinal subsystems are isomorphic; such a category was first defined by Grothendieck in [Gro - 1] and is described in detail in the appendix of [A - M].

(2.1.1) Definitions. A non-empty category J is said to be left filtering if the following holds.

a) Every pair of objects j, j' in J can be embedded in a diagram

$$\begin{array}{ccc} & & j' \\ & \nearrow & \\ j'' & & \\ & \searrow & \\ & & j \ ; \end{array}$$

b) If $j' \rightrightarrows j$ is a pair of maps in J, then there is a map $j'' \longrightarrow j'$ in J such that the compositions $j'' \rightrightarrows j$ are equal.

If C and J are categories, then a J-diagram over C is just a functor $X: J \longrightarrow C$. We will usually use the notation $\{X_j\}$, suppressing both J and the bonding morphisms $X(j) \xrightarrow{X(F)} X(j')$.

A pro - object over C is a J-diagram over C where J is a small left filtering category. The pro - objects over C form a category pro - C with maps defined by

$$(\text{pro} - C)(\{X_j\},\{Y_k\}) \equiv \lim_k \operatorname{colim}_j \{C(X_j, Y_k)\}.$$

(Note: the indexing categories are not assumed equal.)

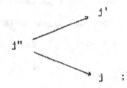

Let tow-C be the full subcategory of pro-C consisting of those objects indexed by the natural numbers. Objects of tow-C are called <u>towers</u>.

We have defined the set of maps in pro-C from X to Y, but the above definition is somewhat opaque and it is not obvious how to define composition of maps from the above definition. Hence, we shall give an alternate definition. For simplicity, consider inverse systems $\{X_j\}$ and $\{Y_k\}$ over C indexed by directed sets J and K respectively.

(2.1.2) <u>Definition</u>. A morphism $f:X \longrightarrow Y$ in pro-C is represented by a map $\theta:K \longrightarrow J$ (not necessarily order-preserving) and morphisms $f_k:X_{\theta(k)} \longrightarrow Y_k$ in C for each k in K such that if $k \leq k'$ in K, then for some j in J with $j \geq \theta(k)$ and $j \geq \theta(k')$ the diagram commutes. Two pairs

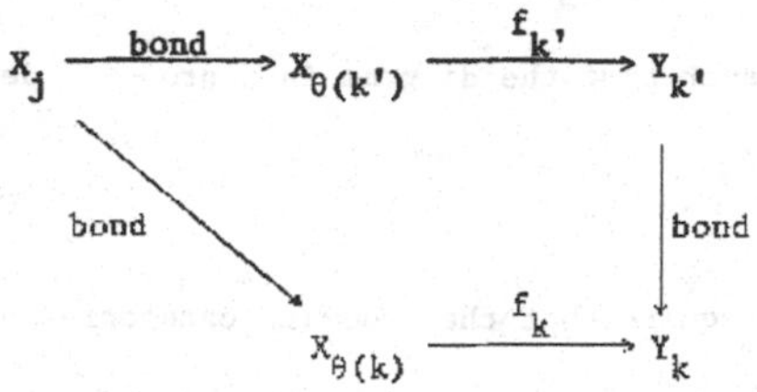

$(\theta,\{f_k\})$ and $(\theta',\{f'_k\})$ represent the same morphism in pro-C if for each k in K there is a j in J with $j \geq \theta(k)$ and $j \geq \theta'(k)$ such that the two composite maps $f_k \circ \text{bond}$, $f'_k \circ \text{bond} : X_j \longrightarrow Y_k$ are equal.

(2.1.3) <u>Remarks</u>. The inverse limit functors $\lim:C^J \longrightarrow C$, if they exist, factor through pro-C. A pro-object $\{X_j\}$ in pro-C contains much more information than its inverse limit $\lim_j\{X_j\}$ in C; the latter need not even exist. The relationship between the pro-object $\{X_j\}$ and its inverse limit

$\lim_j \{X_j\}$ is analogous to the relationship between the germ of a function at a point p and its value at p.

We shall need the following reindexing results from Artin and Mazur [A-M].

(2.1.4) Proposition. A map $f: X \to Y \in$ pro-C can be represented up to isomorphism (in Maps (pro-C)) by a small left filtering inverse system of maps $\{X_j \xrightarrow{f_j} Y_j\}$, i.e., by a pro-object over Maps (C).

More generally, the following holds.

(2.1.5) Proposition. Let Δ be a finite diagram with commutation relations, and suppose that Δ has no loops, i.e., that the beginning and end of a chain of arrows are always distinct. Let D be a diagram in pro-C of the type of Δ, i.e., a morphism of Δ to pro-C. There is a left filtering inverse system $\{D_j\}$ of diagrams of C such that the diagram in pro-C determined by $\{D_j\}$ is isomorphic to D.

Our techniques often require that the indexing categories be cofinite (each element has finitely many predecessors) strongly directed sets ($a \leq b$ and $b \leq a$ implies $a = b$). The following reindexing trick was inspired by Mardesic [Mar-1].

Let I be a small left filtering category and $D \to I$ a diagram over I. An object $d \in D$ will be called an initial object if D contains no maps into d, and for each d' in D, there is a unique map $d \to d'$ in D. Initial objects, if they exist, are clearly unique. Let M(I) be the set of finite diagrams with initial objects over I. We shall call $D \leq D'$ in M(I) if D is a subdiagram of D'. $M(I)$ is clearly cofinite. Because I is filtering, $M(I)$ is a directed set. Further, $D \leq D'$ and $D' \leq D$ implies $D = D'$ (and the initial objects of D and D' are equal). Hence, $M(I)$ is a cofinite strongly directed set.

Define a functor $\text{init}: M(I) \longrightarrow I$ as follows. Associate to a diagram D in $M(I)$ its initial object in I. If $D \leq D'$, there is a unique map in D' from the initial object of D' to the initial object of D. This yields the required functor. Further, the functor init is clearly cofinal.

We thus obtain a functor

$$M: \text{pro}-C \longrightarrow \text{pro}-C,$$

with

$$M\{X_i\}_{i \in I} = \{X_{\text{init}(D)}\}_{D \in M(I)},$$

and a natural equivalence

$$\text{init}: X \longrightarrow M(X).$$

Summarizing, we have the following theorem.

(2.1.6) <u>Theorem</u>. There exists a functor $M: \text{pro}-C \longrightarrow \text{pro}-C$, naturally equivalent to the identity, such that $M(X)$ is indexed by a cofinite strongly directed set for every X in $\text{pro}-C$. □

(2.1.7) <u>Definitions</u>. An object $\{X_j\}$ of $\text{pro}-C$ is called <u>stable</u> if it is isomorphic in $\text{pro}-C$ to an object of C. $\{X_j\}$ is called <u>moveable</u> if for each j there exists a $k > j$ such that for each $\ell > k$ there exists a filler in the diagram

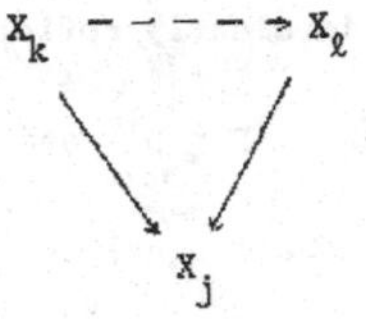

The above description of pro - C may be dualized to yield a category inj - C of <u>direct systems</u> over C in which cofinal systems are isomorphic. Morphisms in inj - C are given by the formula

$$\text{inj} - C\ (\{X_j\},\ \{Y_k\}) = \lim_j\ \text{colim}_k\ \{C(X_j, Y_k)\}.$$

As with pro - C, we shall give an alternative description of a morphism from $\{X_j\}$ to $\{Y_k\}$ in the case that the indexing categories $J = \{j\}$ and $K = \{k\}$ are directed sets. In this case a map may be represented by a function $\theta : J \longrightarrow K$ and maps $f_j : X_j \longrightarrow Y_{\theta(j)}$ in C for each j in J such that if $j < j'$ there exists a k with $k > \theta(j)$ and $k > \theta(j')$ such that the diagram commutes.

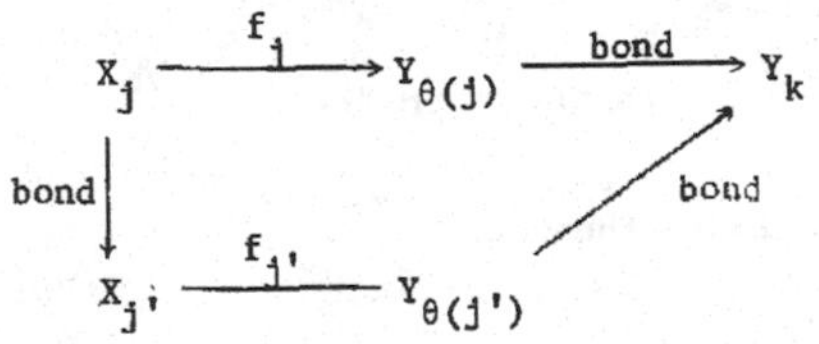

All of the above theory of pro - C, including Artin-Mazur reindexing and the Mardesic construction, may be dualized to inj - C. The colimit functors, colim: $C^J \longrightarrow C$, if they exist, factor through inj - C.

(2.1.8) <u>Proposition</u>. Suppose that C admits an internal mapping functor HOM: $C^{op} \times C \longrightarrow C$. Then HOM extends to a functor HOM: $(\text{pro} - C)^{op} \times C \longrightarrow \text{inj} - C$.

Proof. The required functor is given by

$$HOM (\{X_j\},Y) = \{HOM (X_j,Y)\}. \quad \square$$

§2.2. Some useful categories.

This expository section is intended as a reference for later sections. We suggest that the reader omit the proofs the first time through, and refer to this section as needed later.

We shall give brief sketches of the following categories used in the remainder of these notes.

SS, the category of simplicial sets introduced by D. M. Kan [Kan - 2] and J. C. Moore. See [May - 1] and [Q - 1, §II.3]. D. Quillen's closed model category [Q - 1], a category with sufficient structure to "do homotopy theory," is an abstraction of SS.

CW prespectra and CW spectra (CWSp) introduced by J. Boardman. We follow J. F. Adams [Adams - 1] for the "additive structure" and [Has - 3 - 5] for smash products.

Sp, the category of simplicial spectra developed by D. M. Kan [Kan - 1]. K. S. Brown [Brown] sketched a proof that Sp is a closed model category.

We introduce both categories of spectra because homology theories (which involve smash products) are most easily described in CW spectra, while the closed model structure of simplicial spectra is needed to define a suitable homotopy category of pro - spectra.

M. Tierney [Tier] described stable realization and singular functors inducing an equivalence of homotopy categories Ho(Sp) ~ Ho(CWSp). D. W. Anderson [An - 1,2] gave a different construction of homology theories using chain functors. J. P. May [May - 2] introduced a category of spectra based on the Boardman-Vogt theory of infinite loop spaces [B - V] and thus described the higher structure of ring spectra.

While these approaches are more powerful, the "classical" approach we shall follow is simpler and adequate for our purposes.

We shall then describe how the use of spectra yields a unified treatment of generalized homology and cohomology theories, Spanier-Whitehead duality, and homology and cohomology operations, following [Adams - 1, 3].

We begin by defining simplicial sets.

(2.2.1) Definition. A simplicial set X consists of:

a) A sequence $\{X_n, n \geq 0\}$ of sets. The elements of X_n are called the n-simplicies of X.

b) Face maps $d_i^n: X_n \longrightarrow X_{n-1}$ for $n \geq 1$ and $0 \leq i \leq n$.

c) Degeneracy Maps $s_i^n: X_n \longrightarrow X_{n+1}$ for $n \geq 0$ and $0 \leq i \leq n$.

The maps are required to satisfy the following identities:

$$d_i d_j = d_{j-1} d_i \quad \text{for } i < j,$$

$$s_i s_j = s_{j+1} s_i \quad \text{for } i \leq j,$$

$$d_i s_j = \begin{cases} s_{j-1} d_i & \text{for } i < m, \\ id & \text{for } i = j \text{ or } j+1, \\ s_j d_{i-1} & \text{for } i > j+1. \end{cases}$$

For example, the singular complex S(X) of a topological space X is a simplicial set with typical n-simplex a continuous map $f: \Delta^n \longrightarrow X$, where Δ^n is the standard n-simplex in R^{n+1}.

(2.2.2) Definition. A map $f:X \longrightarrow Y$ of simplicial sets consists of a sequence of maps $f_n:X_n \longrightarrow Y_n$, $n \geq 0$, which satisfy the identities $f_{n-1}d_i = d_if_n$ and $s_if_n = f_{n+1}s_j$.

Definitions (2.2.1) and (2.2.2) combine to yield the category SS of simplicial sets. The following Kan extension condition is crucial to the development of the homotopy theory of SS. Let x be an n-simplex of a simplicial set X. The faces of x, $d_0x, d_1x, \cdots, d_nx$, satisfy certain compatibility conditions, for example, $d_j(d_jx) = d_j(d_{j+1}x)$. A simplicial set X is said to satisfy the Kan extension condition (or is simply called a Kan complex) if given any n $(n-1)$-simplices $y_0, y_1, \cdots, \hat{y}_i, \cdots, y_n$ which satisfy the compatibility conditions to be the faces of an n-simplex, there exists an n-simplex x with $d_jx = y_j$ for $j = 0,1,\cdots,\hat{i},\cdots,n$. More generally, a map $p:E \longrightarrow B$ in SS is called a Kan fibration if given an n-simplex b in B, and n $(n-1)$-simplices in E, $y_0, y_1, \cdots, \hat{y}_i, \cdots, y_n$, which satisfy the appropriate compatibility conditions and the requirement $p(y_j) = d_jb$ for $j = 0,1,\cdots,\hat{i},\cdots,n$, there is an n-simplex x in E with $d_jx = y_j$ for $j = 0,1,\cdots,\hat{i},\cdots,n$, and $p(x) = b$.

For example, the singular complex of a space is a Kan complex; also a Hurewicz fibration $p:E \longrightarrow B$ of spaces induces a Kan fibration $S(p):S(E) \longrightarrow S(B)$.

D. M. Kan [Kan-2] gave a combinatorial description of the homotopy groups of Kan complexes and also described a functorial Postnikov decomposition of Kan complexes. This definition of the homotopy groups is extended to all simplicial sets by defining $\pi_i(X) \equiv \pi_i(SRX)$ $(= \pi_i(RX))$, where $R:SS \longrightarrow Top$ denotes Milnor's [Mil-2] geometric realization functor.

Call a map $f:X \longrightarrow Y$ of simplicial sets a <u>cofibration</u> if it is an inclusion (i.e., each f_n is an inclusion), a <u>weak</u> equivalence if $\pi_0(f)$ is a bijection, and for every choice of basepoints in X, $\pi_*(f)$ is an isomorphism. The usual homotopy-extension and covering-homotopy properties are combined in the following theorem.

(2.2.3) <u>Covering Homotopy Extension Theorem</u>. Given a commutative solid-arrow diagram

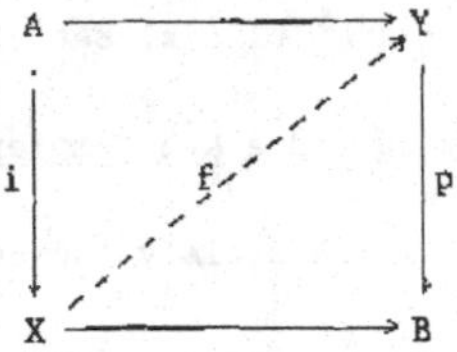

in which i is a cofibration, p is a fibration, and either i or p is a weak equivalence, then there exists a filler f .

See, e.g. [Q - 1, §II.3] for a proof. Theorem (2.2.3) becomes Quillen's Axiom M1 for a model category [Q -1]; see §2.3 .

Note that the usual homotopy extension property <u>only</u> holds for maps into Kan complexes.

(2.2.4) <u>Definitions</u>. A simplicial set X is called <u>finite</u> if X has only finitely many <u>non-degenerate</u> simplices; X is said to have dimension $\leq n$ if X has no non-degenerate simplices in dimensions greater than n.

In §2.4 we shall discuss function spaces in SS .

We shall now sketch the basic properties of CW prespectra and CW spectra. We shall follow J. F. Adams [Adams -1], except for a technical modification made in [Has -3] to construct smash products, for the "additive structure" and follow [Has -3 -5] for smash products and function spectra. Give [0,1] the CW structure with two vertices, namely 0 and 1. Let $S^1 = [0,1]/0 \sim 1$, with the evident basepoint. Let $S^n = S^1 \wedge S^1 \wedge \cdots \wedge S^1$ (n factors), where $K \wedge L = K \times L/K \vee L$.

(2.2.5) <u>Definitions</u>. A <u>CW spectrum</u> $X = \{X_n\}$ consists of a sequence of pointed CW complexes $\{X_n | n = 0,1,2,\cdots\}$, together with <u>cellular inclusions</u> $X_n \wedge S^4 \longrightarrow X_{n+1}$ for each n. A <u>prespectrum map</u> $f:X \longrightarrow Y$ consists of a sequence of continuous pointed maps $f_n:X_n \longrightarrow Y_n$ such that f_{n+1} extends $f_n \wedge id_{S^4}$. The category CWPs of <u>CW prespectra</u> is the category whose objects are CW spectra and whose morphisms are prespectrum maps.

A <u>weak spectrum</u> $X = \{X_n\}$ is a sequence of pointed compactly generated spaces, together with continuous pointed maps $X_n \wedge S^4 \longrightarrow X_{n+1}$ for each n. A <u>weak prespectrum map</u> $f:X \longrightarrow Y$ consists of a sequence of continuous pointed maps $f_n:X_n \longrightarrow Y_n$ such that f_{n+1} extends $f_n \wedge id_{S^4}$ up to homotopy.

The category CWSp of <u>CW spectra</u> is obtained from CWPs by inverting <u>cofinal inclusions</u> of spectra.

(2.2.6) <u>Definitions</u> [Adams -1]. A subspectrum X' of a CW spectrum X is called <u>cofinal</u> if for each cell $\sigma \subset X_n \subset X$ there exists a k such that $(\sigma \cup \{\text{basepoint}\}) \wedge S^{4k} \in X'_{n+k}$. The category of CW spectra, CWSp, is CWPs/{cofinal inclusions}.

(2.2.7) Remarks.

a) The class of cofinal inclusions admits a calculus of right-fractions in the sense of P. Gabriel and M. Zisman [G - Z]. This means that any map $f:X \to Y$ of CW spectra can be represented by a diagram

$$X \supset X' \xrightarrow{f'} Y$$

where X' is cofinal in X and f' is a prespectrum map. Two such diagrams $X \supset X' \xrightarrow{f'} Y$ and $X \supset X'' \xrightarrow{f''} Y$ represent the same map if $f' = f''$ on $X' \cap X''$. Composition is defined as follows. Consider a solid-arrow diagram

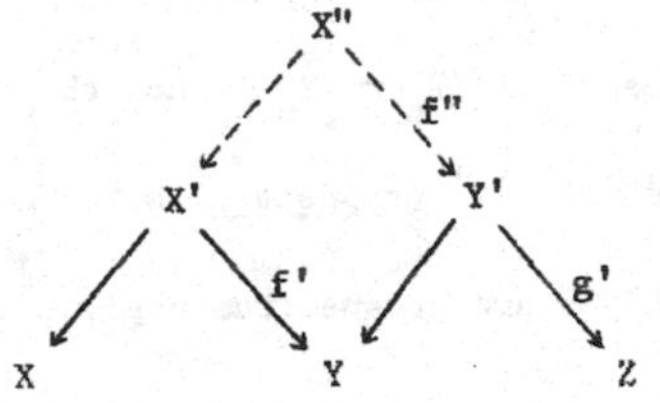

Define a subspectrum $X'' \subset X'$ as follows: X'' consists of those cells of X' which f' maps into Y'. By construction, X'' is cofinal in X', hence cofinal in X. Let $f'' = f'|X''$. The composite of the maps $X \supset X' \xrightarrow{f'} Y$ and $Y \supset Y' \xrightarrow{g'} Z$ is given by $X \supset X'' \xrightarrow{g'f''} Z$. This yields a well-defined composition and makes CWSp a category.

b) Adams uses S^1 where we use S^4 in the definition of CW spectrum. It is easy to see that both definitions yield equivalent categories of CW spectra.

c) By applying realization and cylinder functors and a suitable mapping cylinder functor to a weak spectrum X we obtain a CW spectrum X'

and a natural prespectrum map $X' \longrightarrow X$ which is a weak homotopy equivalence on each level.

(2.2.7) <u>Some examples of CW spectra</u>.

a) <u>The k-sphere spectrum</u> $(k \in Z)$ is given by

$$S^k = \{S^k_n\} = \left\{\begin{matrix} *, & k+4n < 0 \\ S^{k+4n}(\text{space}), & k+4n \geq 0 \end{matrix}\right\},$$

together with the inclusions

$$S^{k+4n}(\text{space}) \wedge S^4\ (\text{space})$$
$$\cong (S^1 \wedge \cdots \wedge S^1) \wedge (S^1 \wedge \cdots \wedge S^1)$$
$$= S^1 \wedge \cdots \wedge S^1$$
$$= S^{k+4n}\ .$$

b) More generally, let K be a pointed CW complex. Associate to K the spectrum

$$K\ (\text{spectrum}) = K\ (\text{space}) \wedge S^0\ (\text{spectrum})$$
$$= \{K \wedge S^{4n}\}\ ,$$

together with the evident inclusions. More generally, we may form $K \wedge S^k$ for any $k \in Z$ (see (2.2.8)). These spectra are called <u>suspension spectra</u>; for sufficiently large n, $(K \wedge S^k)_n \wedge S^4 = (K \wedge S^k)_{n+1}$. If K is finite, we call $K \wedge S^k$ a <u>stable finite CW complex</u>. E. Spanier and J. H. C. Whitehead [S-W] first studied stable finite complexes.

c) The Eilenberg-MacLane spectra. Let $K(G)$ be the CW spectrum associated with the weak spectrum $\{K(G,4n)\}$, together with maps $K(G,4n) \wedge S^4 \longrightarrow K(G,4n+4)$ which classify the fundamental reduced cohomology class. Here G is any abelian group. The Eilenberg-MacLane spectra are examples of Ω-spectra. A CW spectrum X is called an Ω-spectrum if the maps $X_n \longrightarrow \Omega^4 X_{n+1}$ (adjoint to the maps $X_n \wedge S^4 \longrightarrow X_{n+1}$ in X) are homotopy equivalences.

d) The BU-spectrum: $BU_n \simeq BU$ for all n, and the maps $BU_n \wedge S^4 \longrightarrow BU_{n+1}$ are defined by composing Bott periodicity maps $BU \wedge S^2 \longrightarrow BU$. Because the adjoints $BU \longrightarrow \Omega^2 BU$ are homotopy equivalences, BU is an Ω-spectrum. There are similar spectra BO, BSO, BSp, and BSpin associated with the respective "infinite Lie groups." BU classifies complex K-theory, BO real K-theory, etc., see (f) below.

e) The Thom spectra MU, MO, MSO, MSp, M spin. Consider the classifying bundles $EU(m) \longrightarrow BU(m)$. The Thom complex (see, e.g. [Sto]) $MU(m)$ is the quotient of the unit disk bundle $DU(m)$ in $EU(m)$ by the unit sphere bundle $Sph\ U(m)$ in $EU(m)$. Note that $MU(1) \cong D^2/S^1 \cong S^2$. There are Whitney sum maps

$$\begin{aligned} MU(\ell) \wedge MU(m) &= DU(\ell) \times DU(m) / Sph\ U(\ell) \times Sph\ U(m) \\ &\longrightarrow DU(\ell+m) / Sph\ U(\ell+m) \\ &= MU(\ell+m). \end{aligned}$$

These yield inclusions

$$MU(m) \wedge S^2 \longrightarrow MU(m+1),$$

and thus the MU spectrum $MU = \{MU(2n)\}$. Similar constructions yield the other Thom spectra. MU classifies complex cobordism by a classical result of Thom, see e.g., [Sto]. Similarly, the other Thom spectra classify appropriate cobordism theories. Cobordism is important because: (1) some varieties are quite powerful; and (2) the dual homology theories, bordism, have natural geometric definitions. R. Stong [Sto] is a good source on cobordism.

f) The examples (c), (d), and (e) above classify well-known cohomology theories: $H^*(;G)$, various forms of K-theory, and various forms of cobordism theory respectively. Brown's theorem associates to any generalized cohomology theory E^* a spectrum $E = \{E_n\}$ with $E^{4n}(X) \cong [X, E_n]$.

(2.2.8) Definition. If K is a pointed CW complex and X is a CW spectrum, then their smash product is $K \wedge X = \{K \wedge X_n\}$, together with inclusions $K \wedge X_n \wedge S^4 \longrightarrow K \wedge X_{n+1}$ induced from X.

(2.2.9) Definition. Maps $f,g:X \rightrightarrows Y$ of CW spectra are homotopic if there is a map $H:([0,1] \cup *) \wedge X \longrightarrow Y$ such that $H|(0 \cup *) \wedge X = f$ and $H|(1 \cup *) \wedge X = g$.

Let Ho(CWSp) denote the resulting homotopy category of CW spectra. Ho(CWSp) is essentially the category introduced by Boardman (see [Vogt-2]) as formulated by Adams [Adams-1]. Caution: CWSp is not a model category.

The homotopy groups of a CW spectrum are given by $\pi_k X = \mathrm{Ho}(CWSp)(S^k, X)$ for $k \in Z$.

(2.2.10) Remarks. The homotopy category of stable finite CW complexes, a full subcategory of Ho(CWSp), is the classical Spanier-Whitehead category (use the Freudenthal suspension theorem which states that $[X \wedge S^n, Y \wedge S^n] \cong [X \wedge S^{n+1}, Y \wedge S^{n+1}]$ for finite complexes X and Y and sufficiently large n).

More generally call a CW spectrum X finite if for some N, X_N is a finite complex and $X_n = X_N \wedge S^{4n-4N}$ for $n \geq N$. If X and Y are finite, and X_N and Y_M are the above complexes, then

$$\text{(2.2.11a)} \qquad CWSp(X,Y) \cong \mathrm{colim}_k \{CW(X_N \wedge S^{k-4N}, Y_M \wedge S^{k-4M})\},$$

and

$$\text{(2.2.11b)} \qquad \mathrm{Ho}(CWSp)(X,Y) \cong \mathrm{colim}_k \{[X_N \wedge S^{k-4N}, Y_M \wedge S^{k-4M}]\}.$$

The category CWSp is then defined so that a spectrum is the colimit of its finite subspectra. In fact, a CW spectrum is the homotopy colimit (see §4.10) of its finite subspectra. J. Boardman (see [Vogt -2]) and A. Heller [Hel -1,2] define morphisms of CW spectra by this criterion and (2.2.11a).

Cofibrations and homotopy equivalences behave similarly in CW spectra and CW complexes. A. Heller formalized these properties by introducing abstract h-c categories [Hel -1,2].

We shall need smash products and function spectra in order to discuss generalized homology and cohomology theories. These constructions are more difficult than the above structure. In fact, there is no smash product on Ho(CWSp) with S^0 as unit [Has-5]. J. Boardman (see [Vogt 2]) gave the first construction of a coherently homotopy associative, commutative, and unitary (S^0) smash product for CW spectra. D. M. Kan and G. W. Whitehead [K-W] described a non-associative smash product for simplicial spectra. Later Adams [Adams-1] and still later the second author [Has-3] gave different and simpler constructions for the smash product of CW spectra. J. P. May [May-2] and D. Puppe [Puppe] gave a radically different construction following the Boardman-Vogt theory of infinite loop spaces [B-V]. We shall follow [Has-3].

(2.2.12) The interchange problem. We may regard the sphere spectrum S^0 as a (non-commutative) ring and any CW spectrum as a right-module over $S^0 = \{S^{4n}\}$. Construction of a smash product requires permutations π of $S^4 \wedge \cdots \wedge S^4$. Because S^0 is only homotopy commutative, this requires canonical homotopies H_π from π to the identity. These are defined as follows. Identify

$$S^{4k} \cong S^4 \wedge \cdots \wedge S^4 \cong C^{2*} \wedge \cdots \wedge C^{2*}$$
$$\cong (C^2 \times \cdots \times C^2)^*$$

where $*$ denotes the one-point compactification. Then π simply permutes factors of $C^2 \times \cdots \times C^2$; thus $\pi \in SU(2k)$. Because $SU(2k)$ is path connected and simply connected, there is a unique homotopy class of paths $[\Gamma_\pi]$ in $SU(2k)$ with $\Gamma_\pi(0) = \pi$ and $\Gamma_\pi(1) = e$, the identity of $SU(2k)$. Define

$$H_\pi:([0,1] \cup *) \wedge S^{4k} \cong [0,1] \times S^{4k} \longrightarrow S^{4k}$$

by

$$H_\pi(t,x) = \Gamma_\pi(t)(x).$$

Then $[H_\pi]$ is the required homotopy class (relative to the endpoints) of canonical homotopies.

We shall now define a family of smash products on CWSp, and prove that they are all equivalent and have the required properties.

(2.2.13) Definition. Given a sequence of pairs of non-negative integers

$$\{(i_n, j_n) \mid n \geqq 0,\ i_n + i_n = n, \text{ and } \{i_n\}$$
$$\text{and } \{j_n\} \text{ are monotone unbounded sequences}\},$$

define an associated smash product by

$$X \wedge Y = \{(X \wedge Y)_n\} = \{X_{i_n} \wedge Y_{j_n}\}$$

together with the appropriate inclusions induced from X and Y.

Then $\wedge$ extends to bifunctors on CWPs, CWSp (the smash product of cofinal inclusions is a cofinal inclusion), and Ho(CWSp).

(2.2.14) Theorem. Any two sequences (2.2.13) yield canonically equivalent smash products on Ho(CWSp).

We shall need the following machinery.

(2.2.15) Definition. Let X be a CW spectrum. Given a monotone unbounded sequence of non-negative integers

$$\{j_n \mid n \geqq 0,\ j_n \leqq n\},$$

define a CW spectrum

$$DX = \{(DX)_n\} = \{X_{j_n} \wedge S^{4n-4j_n}\}$$

together with the appropriate inclusions induced from X.

Then D extends to a functors (called destabilization) on CWSp. There are natural cofinal inclusions $DX \longrightarrow X$, so that D is naturally equivalent to the identity.

(2.2.16) Definition. Let X be a CW prespectrum. A permutation π of DX consists of a sequence of maps

$$\{\pi_n = id \wedge \pi_n : (DX)_n = X_{j_n} \wedge S^4 \wedge \cdots \wedge S^4$$

$$\longrightarrow Y_{j_n} \wedge S^4 \wedge \cdots \wedge S^4$$

$$= (DX)_n\}$$

where each π_n is a permutation of $S^4 \wedge \cdots \wedge S^4$. A sequence of maps

$$g = \{g_n : (DX)_n \longrightarrow Y_n | g_{n+1}$$

$$\text{extends } g_n \text{ up to permutation}\}$$

is called a permutation map.

(2.2.17) Proposition. Permutation maps are weak maps, where the required homotopies $H_n:(DX)_n \wedge S^4 \times [0,1] \longrightarrow Y_{n+1}$ are induced from canonical homotopies (2.2.12).

The proof is easy and omitted.

The following lemmas relating weak maps and maps of CW spectra are proved in [Has -3]. The proofs involve construction of suitable mapping cylinders and the homotopy extension property.

(2.2.18) Lemma. A weak map $f:X \longrightarrow Y$, together with a family of homotopies for f,

$$\{H_n:X_n \wedge S^4 \times [0,1] \longrightarrow Y_{n+1}\}$$

(see (2.2.5)), induces a strict map $F:S \to Y$.

F depends upon $\{f\}$ and $\{H_n\}$ up to equivalence in $Ho(CWSp)$.

(2.2.19) <u>Lemma</u>. Let $\{H_n\}$ and $\{H'_n\}$ be families of homotopies for a weak map f. If $H_n \simeq H'_n$ relative to the endpoints for each n, then $(f,\{H_n\})$ and $(f,\{H'_n\})$ induce homotopic maps $X \to Y$ in $CWSp$.

(2.2.20) <u>Lemma</u>. Let $f':X \to Y$ and $f'':Y \to Z$ be weak maps with homotopies $\{H'_n\}$ and $\{H''_n\}$ respectively. Define "composed" homotopies

$$H_n : X_n \wedge S^4 \times [0,1] \longrightarrow Z_{n+1}$$

by

$$H_n(x,s,t) = \begin{cases} H''_n\big(f'_n(x),s,2t\big), & 0 \le t \le \tfrac{1}{2} \\ f''_{n+1}\big(H_n(x,s,2t-1)\big), & \tfrac{1}{2} \le t \le 1. \end{cases}$$

Let F', F'', and F be the maps associated with $(f',\{H'_n\})$, $(f'',\{H''_n\})$, and $(f''f',\{H_n\})$. Then $F \simeq F'F''$ in $Ho(CWSp)$.

Lemmas (2.2.18) - (2.2.20) yield the following proposition.

(2.2.21) <u>Proposition</u>. A permutation-commutative diagram of permutation maps induces a commutative diagram in $Ho(CWSp)$. The maps and required homotopies in this diagram are defined up to homotopy in $Ho(CWSp)$. □

(2.2.22) <u>Proof of Theorem (2.2.14)</u>. There are destabilizations (2.2.15) and natural permutation classes of permutation maps $D(X \wedge Y) \to X \wedge' Y$, $D'(X \wedge' Y) \to X \wedge Y$. The composite mappings $D'D(X \wedge Y) \to X \wedge Y$ and

$DD'(X \wedge' Y) \longrightarrow X \wedge' Y$ differ from the respective (cofinal) inclusions by permutations. The conclusion follows. □

(2.2.23) Theorem. There are natural maps in Ho(CWSp),

$$X \longrightarrow X \wedge S^0 \longrightarrow X \qquad \text{(unit)},$$

$$a:(X \wedge Y) \wedge Z \longrightarrow X \wedge (Y \wedge Z) \qquad \text{(associativity)},$$

$$c:X \wedge Y \longrightarrow Y \wedge X \qquad \text{(commutativity)},$$

which yield a symmetric monoidal category in the sense of S. Eilenberg and G. M. Kelly [E-K].

Proof. There are destabilizations and natural permutation classes of permutation maps

$$DX \longrightarrow X \wedge S^0 \longrightarrow X,$$

$$a':D((X \wedge Y) \wedge Z) \longrightarrow X \wedge (Y \wedge Z),$$

$$c':D(X \wedge Y) \longrightarrow Y \wedge X$$

(D is used generically). Let the maps $DX \longrightarrow X \wedge S^0 \longrightarrow X$, a, and c be the associated maps of CW spectra ((2.2.17) - (2.2.18)).

By Proposition (2.2.21), it suffices to obtain the coherency diagrams [E-K] as permutation-commutative diagrams of permutation maps between destabilizations. These diagrams include statements that the above maps are isomorphisms, that the composite map $c^2:D^2(X \wedge Y) \longrightarrow X \wedge Y$ is homotopic to the identity, and that certain coherency conditions hold. For example, the diagram stating that $\wedge$ is coherently homotopy associative is

$$\begin{array}{ccc}
D^3(((W \wedge X) \wedge Y) \wedge Z) & \xrightarrow{D^2(a')} & D^2((W \wedge X) \wedge (Y \wedge Z)) \\
\big\downarrow (a' \wedge id)_* & & \big\downarrow D(a') \\
D^2((W \wedge (X \wedge Y)) \wedge Z) & & D(W \wedge (X \wedge (Y \wedge Z))) \\
\big\downarrow D(a') & & \big\downarrow \\
D(W \wedge ((X \wedge Y) \wedge Z) & \xrightarrow{(id \wedge a')_*} & W \wedge (X \wedge (Y \wedge Z)).
\end{array}$$

These diagrams are readily obtained. □

(2.2.24) Remarks. The approach of Theorems (2.2.22) - (2.2.23) yields a suitable smashproduct on Ho(CWSp) for studying homology and cohomology theories and operations involving maps of spectra. It does however ignore higher homotopies which yield the rich structure of infinite loop spaces [May - 3]. May and Puppe incorporate this structure in their smash products. It would be interesting to obtain higher homotopies within the above simple framework, perhaps with a suitable operad. See P. Malraison [Mal].

(2.2.25) Further properties of $\wedge$. It is easy to verify that $\wedge$ commutes with the suspension

$$S^1 \wedge ?: Ho(CWSp) \longrightarrow Ho(CWSp),$$

$\wedge$ satisfies a Kunneth formula for stable integral homology. and $\wedge$ is weakly universal for pairings [Wh - 1], [K - W].

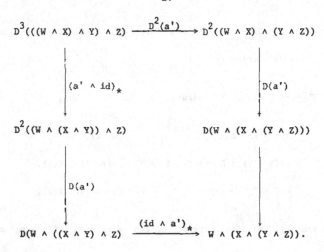

We shall now define function spectra, and verify that Ho(CWSp), together with the smash product $\wedge$ and function spectra, forms a symmetric monoidal closed category in the sense of Eilenberg and Kelly [E - K]. This includes the usual exponential law.

(2.2.26) Construction of function spectra. Let X and Y be CW spectra. For a first approximation to the function spectrum $HOM(X,Y)$, define a weak spectrum $Map(X,Y)$ as follows. Let S^{-4k} be the $-4k$-sphere spectrum: $(S^{-4k})_n = S^{4n-4k}$ for $n \geq k$, and $*$ otherwise; the required inclusions are induced by isomorphisms $S^{4n-4k} \wedge S^4 \cong S^{4n-4k+4}$. Choose a representative smash product $\wedge'$ on CWSp using (2.2.13). The eventual function spectrum in Ho(CWSp) will be independent of this choice. Define

$$Map(X,Y)_n = CWSp(S^{-4n} \wedge X, Y)$$

with the topology induced from the compactly generated function spaces $Map((S^{-4n} \wedge X')_i, Y_i)$ where X' is a cofinal subspectrum of X. The maps $S^4 \wedge S^{-4n-4} \longrightarrow S^{-4n}$ (S^4 is a space, the other terms are spectra) induce the required maps

$$Map(X,Y)_n \wedge S^4 \longrightarrow Map(X,Y)_{n+1}.$$

This yields a weak spectrum $Map(X,Y)$.

(2.2.27) Definition. Let $HOM(X,Y)$ be the CW spectrum obtained from $Map(X,Y)$, see Remarks (2.2.7)(c).

We can extend HOM to bifunctors on CWSp and Ho(CWSp). We shall now show that HOM is the required internal mapping functor.

(2.2.28) Theorem (exponential law).

$$Ho(CWSp)(X \wedge Y,Z) \cong Ho(CWSp)(X, HOM(Y,Z)).$$

<u>Proof</u>. If X and Y are finite spectra, it is easy to show that

$$\text{CWSp}(X \wedge' Y, Z) = \text{CWSp}(X, \text{Map}(Y,Z)). \qquad (2.2.29)$$

Because a CW spectrum is the colimit of its finite subspectra (the Boardman-Heller completion, see (2.2.11) and the following discussion), (2.2.29) holds for arbitrary CW spectra. This also yields an analogous formula in Ho(CWSp). Finally, there are natural weak homotopy equivalences $\text{HOM}(Y,Z) \longrightarrow \text{Map}(Y,Z)$ (from projections of mapping cylinders and the natural transformation $R \circ \text{Sin} \longrightarrow \text{id}$). Thus

$$\text{Ho(CWSp)}(X, \text{Map}(Y,Z)) \cong \text{Ho(CWSp)}(X, \text{HOM}(Y,Z)),$$

by [Adams -1, Theorem 3.4]. The conclusion follows. □

(2.2.30) <u>Corollary</u>. $\text{HOM}(\ ,\ ) = \text{Ho(CWSp)}(\ ,\ )$.

This follows from the definition $\pi_0 = \text{Ho(CWSp)}(S^0,\ )$. This also shows that Ho(CWSp) is <u>normalized</u> [E - K, p. 491].

(2.2.31) <u>Theorem</u>. Ho(CWSp), together with the above structure, is a symmetric monoidal closed category.

The remaining coherence conditions [E - K, p. 491, Theorem 5.5] are easily verified. Their precise statement and proof is omitted.

(2.2.32) <u>Remarks</u>. Adams defines an internal mapping functor HOM by appealing to Brown's Theorem (see [Adams - 1]). This approach also yields the above theorem.

This concludes our formulation of the category of CW spectra. We shall now briefly discuss simplicial spectra and the equivalence of homotopy categories. At the end of this section we shall use CW spectra to discuss homology and cohomology theories and operations, following [Adams - 1,3].

(2.2.33) Definition [Kan - 1]; we follow [May - 1]. The simplicial suspension functor $E:SS_* \longrightarrow SS_*$ is defined on objects as follows. Let X be a pointed simplicial set with basepoint $*$. Then $(EX)_0 = *_0$, and $(EX)_n$ consists of all symbols (i,x) where i is a positive integer and $x \in X_{n-i}$, subject to the identifications $(i, *_{n-i}) = *_n$ ($*_k$ denotes the appropriate degeneracy of the basepoint in X_k). Face and degeneracy maps are defined as follows:

$$s_0(i,x) = (i+1,x)$$

$$s_{i+1}(1,x) = (1,s_i x)$$

$$\partial_0(1,x) = *_n, \quad x \in X_n$$

$$\partial_1(1,x) = *_0, \quad x \in X_0$$

$$\partial_{i+1}(1,x) = (1,\partial_i x), \quad x \in X_n, \quad n > 0,$$

and by the simplicial identities (2.2.1).

Then EX is a pointed simplicial set with one non-degenerate n simplex for every non-degenerate $(n-1)$-simplex of X except the basepoint. Further, the reduced suspension of the realization of X, ΣRX is canonically homeomorphic to the realization REX.

(2.2.34) Definitions [Kan - 1], compare Definitions (2.2.5). A simplicial prespectrum X consists of a sequence of pointed simplicial sets $X(n)$, together with inclusions $EX(n) \longrightarrow X(n+1)$. A map of simplicial prespectra $f:X \longrightarrow Y$ consists of a sequence of pointed simplicial maps $f(n):X(n) \longrightarrow Y(n)$ such that $f(n+1)$ extends $Ef(n)$. Let Ps denote the category of simplicial prespectra.

We could "complete" the category of simplicial prespectra as in Definitions (2.2.6). However, Kan introduced a conceptually simpler completion which essentially replaces a simplicial prespectrum by its associated Ω - spectrum (see (2.2.7)(b)).

An $n+k$ simplex of $X(n)$ (element of $X(n)_{n+k}$) corresponds to an $(n+k+1)$ - simplex of $X(n+1)$ under E. Thus Kan defined the simplicial spectrum SX associated to X by taking as stable k - simplices the pointed (*) sets

$$SX_k = \cup X(n)_{n+k}, \quad k \in Z, \quad n+k \geq 0.$$

There are induced face and degeneracy maps of pointed sets

$$d_i : SX_{k+1} \longrightarrow SX_k$$

$$s_i : SX_k \longrightarrow SX_{k+1}$$

for $i \geq 0$ which satisfy

i) the usual simplicial identities (2.2.1)

and

ii) the local finiteness condition: for every simplex σ in SX, there is an integer n (depending upon σ) such that $d_i\sigma = *$ for $i > n$.

(2.2.35) Definitions [Kan - 1]. A simplicial spectrum X consists of pointed (*) sets X_k (the k - simplices of X), $k \in Z$, together with face and degeneracy maps which satisfy the above conditions. A map of simplicial spectra

$f:X \longrightarrow Y$ consists of a sequence of maps of pointed sets $\{f_k:X_k \longrightarrow Y_k\}$ which commute with face and degeneracy maps. Let Sp be the <u>category of simplicial spectra</u>.

The spectrum construction S above extends to a functor $S:Ps \longrightarrow Sp$. S admits an adjoint <u>prespectrum functor</u> $P:Sp \longrightarrow Ps$, on a simplicial spectrum X, PX is defined by letting $PX(n)_j$ consist of those $(j-n)$ simplices σ of Y with $d_i\sigma = *$ for $i > j$. Further, SP = identity:Sp ⟶ Sp.

The categories Sp and SS enjoy many similar properties. In particular, there is a <u>Kan extension condition</u> for simplicial spectra, and the homotopy groups of a Kan spectrum admit a combinatorial definition [Kan -1]. More generally, K. S. Brown proved the following theorem.

(2.2.36) <u>Theorem</u> [Brown]. The category of simplicial spectra admits a natural closed model structure in the sense of D. G. Quillen [Q -1], see §2.3.

We shall sketch an independent proof in the spirit of Quillen's proof [Q -1, §II.3] that SS is a closed model category. The first task is to define <u>standard simplices</u> which do not exist within Sp, but only in a larger category BSp of <u>big simplicial spectra</u>.

The definition of a <u>big simplicial spectrum</u> and the category BSp is analogous to the definition of simplicial spectra and Sp (2.2.35) except that there is <u>no local finiteness restriction</u> (ii).

Sp is a full subcategory of BSp. Also, the inclusion J:Sp ⟶ BSp admits an adjoint T ; on objects in BSp, TX_k consists of those simplices in X_k with almost all faces at the basepoint (condition (ii) in (2.2.35)). Alternatively, extend the functor $P:Sp \longrightarrow Ps$ to BSp; then $T = SP$.

Following V. K. A. M. Gugenheim [Gug, p. 36], a <u>simplicial operator</u> ϕ is a composite of face and degeneracy maps. Each simplicial operator ϕ has a unique <u>standard representative</u>.

$$\phi = s_{m_i} \cdots s_{m_2} s_{m_1} d_{n_1} d_{n_2} \cdots d_{n_j} ,$$

with $0 \leq m_1 < m_2 < \cdots < m_i$ and $0 \leq n_1 < n_2 < \cdots < n_j$. The <u>height</u> of ϕ is defined to be $i - j$.

(2.2.37) <u>Definition</u>. The standard k-simplex Δ^k is the big simplicial spectrum with one non-degenerate k-simplex δ_k and whose m-simplices are given by

$$\Delta^n_m = \{\phi\delta_n\} \cup \{*\},$$

where ϕ ranges over all simplicial operators of height $m-k$, together with a disjoint basepoint $*$.

(2.2.38) <u>Proposition</u>. The k-simplices of a big simplicial spectrum are in one-to-one correspondence with the maps $\Delta^k \longrightarrow Y$.

<u>Proof</u>. As in [G-Z, Theorem 1]. □

(2.2.39) <u>Remarks</u>. For Y in S_p, the maps $\Delta^k \longrightarrow Y$ satisfy a local finiteness condition analogous to condition (ii).

(2.2.40) <u>Definition</u>. The <u>boundary</u> of Δ^k, $\partial\Delta^k$, is the big simplicial sub-spectrum of Δ^k generated by the faces $d_i\delta_k$ of Δ^k, $i \geq 0$. The <u>horn</u> $V^{k,\ell}$

is the big simplicial subspectrum of $\partial\Delta^k$ generated by the faces $d_i\delta_k$ for $i \neq \ell$.

We discuss the extension condition for BSp and Sp, and the homotopy groups of a simplicial spectrum.

(2.2.41) <u>Definition</u> (Compare [Q-1, II.32 for SS). A big simplicial spectrum X is said to be <u>Kan</u> if every map $V^{k,\ell} \longrightarrow X$ can be extended to Δ^k.

(2.2.42) <u>Remarks</u>. Kan [Kan-1, Def. 7.3] said that a simplicial spectrum X satisfied the extension condition if for each n, the simplicial set PX(n) satisfies the Kan extension condition, see, e.g., [May-1, Def. 1.3]. It is easy to check that the definitions are equivalent.

We shall now define the homotopy groups of a Kan simplicial spectrum, and extend the definition to Sp. Let X be a Kan simplicial spectrum.

(2.2.43) <u>Definition</u>. (Compare, e.g., [May-1, Def. 3.1]). Two k-simplices σ and σ' of X are <u>homotopic</u>, denoted $x \sim x'$, if for all i, $d_i\sigma = d_i\sigma'$, and if there exists a $(k+1)$-simplex τ with $d_0\tau = \sigma$, $d_1\tau = \sigma'$, and $d_i\tau = d_is_0\sigma = d_is_0\sigma'$ for $i \geq 2$.

(2.2.44) <u>Proposition</u>. (Compare [May-1, Proposition 3.2]). If X is a Kan spectrum, then $\sim$ is an equivalence relation on the k-simplices of X, for all k.

<u>Proof</u>. As in the proof of [May-1, Proposition 3.2]. □

(2.2.45) <u>Definition</u>. (Compare [May-1, Def. 3.6]). Let X be a Kan spectrum. Let $\tilde{X}_k$ denote the set of all of the k-simplices σ of X which satisfy $d_i\sigma = *$ for all i. Define $\pi_k(X) = \tilde{X}_k/\sim$.

By imitating the discussion in [May - 1, §4], we see that

(2.2.46) Proposition. $\pi_k(X)$ is an abelian group. □

Finally, for a pointed simplicial set $(K,*)$ which satisfies the extension condition, define $\sim$, $\tilde{K}_m$ and $\pi_m(K)$ as above for $m \geq 0$. $\pi_m(K)$ is a group for $m \geq 1$.

(2.2.47) Proposition. $\pi_k(X) = \pi_{k+n}PX(n)$ for all $k+n \geq 1$.

Proof. A k-simplex σ in $\tilde{X}_k$ can be realized in $PX(n)$ for $n+k \geq 0$ since $d_i\sigma = *$ for all i, hence for $n+k \geq 0$,

$$\tilde{X}_k \subset (PX(n))^{\sim}_k .$$

Similarly, for $n+k \geq 0$,

$$(PX(n))^{\sim} \subset \tilde{X}_k .$$

Further the simplices needed to define the equivalence relation in $\tilde{X}_k$ may be realized in $PX(n+k)$ for $n+k \geq 0$, and conversely. The conclusion follows (we need $n+k \geq 1$ for $\pi_{n+k}PS(n)$ to be a group). □

Hence our definition of $\pi_k(X)$ agrees with that of Kan [Kan - 1, §10].

The following proposition is essentially contained in [Kan - 1, §8-10]. We give an explicit proof because of the variety of equivalent definitions of weak equivalence in SS (the equivalent conditions in [Q - 1, Proposition §II.3.4]).

(2.2.49) Proposition. A map $f: X \to Y$ in Sp is a weak equivalence if and only if for $n \geq 0$, the maps $Pf(n):PX(n) \longrightarrow PY(n)$ are weak equivalences.

<u>Proof</u>. Let $f:X \longrightarrow Y$ be a weak equivalence. We shall show that for any $PX(n)$ and for any basepoint x in $PX(n)$, and for any $m \geq 0$, the induced map

$$Pf(n)_*:\pi_m(PX(n),x) \longrightarrow \pi_n(PY(n),y),$$

where $y = Pf(n)(x)$, is an isomorphism. To see this, since $EPX(n)$ is a connected subsimplicial set of $PX(n+1)$, and contains the standard basepoint $*$ (the basepoint of X),

$$\begin{aligned}\pi_m(PX(n),x) &\cong \pi_{m+1}(PX(n+1),Ex)\\ &\cong \pi_{m+1}(PX(n+1),*)\\ &\cong \pi_{m+1}(PY(n+1),*) \quad \text{by Proposition 4.7)}\\ &\cong \pi_{n+1}(PY(n+1),E_y)\\ &\cong \pi_m(PY(n),y);\end{aligned}$$

further, the isomorphisms induced by changing basepoints may be chosen so that the composite is $Pf(n)_*$. Hence the map on realizations

$$R(Pf(n)) : R(PX(n)) \longrightarrow R(PY)(n))$$

is a weak equivalence. By [Q -1, Proposition §II.3.4], so is the map $Pf(n)$.

Proof of the converse follows immediately from Proposition (2.2.47). □

(2.2.49) Remarks. Relative homotopy groups, the homotopy exact sequence for a pair, the fibre of a fibration, and the homotopy exact sequence of a fibration for Sp, as in [Kan -3, §§2,3], may be obtained by imitating their developments in SS. See, e.g., [May -1, §§4-7].

Cofibrations and fibrations in Sp are defined analogously with the definitions in SS [Kan -2], see e.g., [May -1], [Q -1, §II,3].

(2.2.50) Definitions. Cofibrations are injective maps. A map $E \longrightarrow B$ of simplicial spectra is a (Kan) fibration if given any commutative solid-arrow diagram in BSp of the form

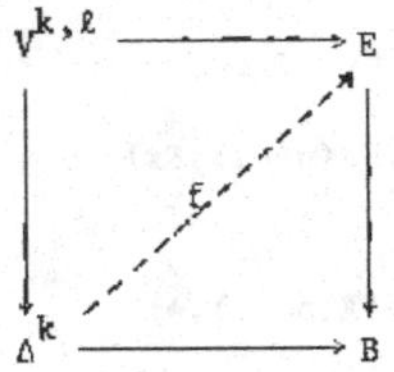

there exists a filler f.

We can now prove Theorem (2.2.36), that Sp is a closed model category, by imitating Quillen's proof that SS is a closed model category [Q -1, §II,3]. Details are omitted.

We shall now summarize Adams' formulation of the foundations of generalized homology and cohomology theories; see [Adams -1,3]. Let h_* be a non-negatively graded generalized homology theory defined on finite complexes and h^* be the dual cohomology theory.

(2.2.51) Brown's Theorem [Bro]. There is a CW spectrum $E = \{E_n\}$ such that $h^{4n}(X) \cong Ho(CW_*)(X, E_{4n})$ for all finite pointed CW complexes X.

(2.2.52) Theorem (G. W. Whitehead [Wh -2]). With h_* and $E = \{E_n\}$ as above,

$$h_k(X) = \operatorname{colim}_{n \to \infty} \mathrm{Ho}(CW)\left(S^{4n+k}, X \wedge E_n\right).$$

We use (2.2.51) and (2.2.52) to define h^* and h_* on all CW <u>spectra</u>:

(2.2.53)
$$\begin{aligned} h_k(X) &\equiv \mathrm{Ho}(CWSp)(S^k, X \wedge E) \\ &\equiv \pi_k(X \wedge E) \\ h^k(X) &\equiv \mathrm{Ho}(CWSp)(X \wedge S^{-k}, E). \end{aligned}$$

(2.2.54) <u>Alexander and Spanier-Whitehead duality</u>. Let K be a compact polyhedron linearly embedded in S^n. By Alexander duality there are isomorphisms

$$H^p(K) \cong H_{n-p-1}(S^n \setminus K).$$

E. Spanier and G. W. Whitehead extended Alexander duality to state that K determines the stable homotopy type of $S^n \setminus K$ and that $S^n \setminus K$ has the homotopy type of a compact polyhedron.

Spanier introduced the following formulation of duality. Let K and L be compact polyhedra disjointly embedded in S^n. Choose a PL path ω from K to L with $\omega(0) \in K$, $\omega(1) \in L$, and $\omega(0,1)$ disjoint from $K \cup L$. Regarding S^n as the compactification of R^n with $\omega(\frac{1}{2})$ as the "point at ∞," yields disjoint embeddings of K and L in R^n. Let

(2.2.55)
$$\mu : K \times L \longrightarrow S^{n-1}$$

be the map

(2.2.56)
$$\mu(k,\ell) = \frac{k-\ell}{\|k-\ell\|}.$$

It is easy to check that the restrictions $\mu|\omega(0) \times L$ and $\mu|K \times \omega(1)$ are null-homotopic. Let $\omega(0)$ and $\omega(1)$ be the basepoints of K and L respectively. Then (2.2.37) yields a map

$$\mu: K \wedge L \longrightarrow S^{n-1}. \qquad (2.2.55)$$

Regarding K, L, and S^{n-1} as spectra, and taking the adjoint of μ in (2.2.57), yields a map

$$\mu_*: K \longrightarrow \mathrm{HOM}\ (L, S^{n-1}). \qquad (2.2.58)$$

Spanier proved that if the inclusion $L \longrightarrow S^n \setminus K$ is a homotopy equivalence, then μ_* is a stable homotopy equivalence. K and L play symmetric roles in (2.2.57) and (2.3.58). $S^{n-1} \setminus K$ is called the $(n-1)$-dual of K:

$$D_{n-1}K \equiv S^{n-1} \setminus K. \qquad (2.2.59)$$

Then, $D_{n-1}D_{n-1} \cong \mathrm{id}$. Finally, the natural "composition"

$$D_{n-1}K \wedge \mathrm{HOM}(S^{n-1}, E) \xrightarrow{\simeq} \mathrm{HOM}\ (K, S^{n-1}) \wedge \mathrm{HOM}\ (S^{n-1}, E) \longrightarrow \mathrm{HOM}\ (K, E) \qquad (2.2.60)$$

is a stable homotopy equivalence.

(2.2.61) Definition. The functional dual of a CW spectrum X is given by

$$DX \equiv \mathrm{HOM}\ (X, S^0).$$

There is a natural map $X \longrightarrow D^2X$ (take the adjoint of the evaluation map $X \wedge DX = S \wedge \mathrm{HOM}\ (X, S^0) \longrightarrow S^0$) which is a stable homotopy equivalence if X is finite.

We shall now discuss generalized homology theories represented by ring spectra, following Adams [Adams - 3, especially pp. 60-68].

(2.2.62) Definition. A ring spectrum consists of a CW spectrum E, a multiplication map $m: E \wedge E \longrightarrow E$, and a unit map $i: S^0 \longrightarrow E$. We require that m be homotopy associative and commutative and that i be a homotopy unit.

J. P. May has gone considerably farther in studying the higher homotopies associated with ring spectra [May -2,3].

(2.2.63) Examples (see (2.2.7) for descriptions).

a) S^0.

b) K(R), where R is a ring.

c) BU, BO, etc; the multiplication is induced from the tensor product of vector bundles.

d) MU, MO, etc.; the multiplication is induced from the Whitney sum of vector bundles.

e) bu, bo, etc.; connected versions of BU, BO, etc.; see D. W. Anderson [An -1,2].

Module spectra are defined analogously with (2.2.62). All spectra are modules over S^0. P. E. Conner and E. E. Floyd [Con - Fl] showed that BU is a module spectrum over MU.

(2.2.64) Let E_* be a generalized homology theory represented by a ring spectrum E, i.e., $E_*(X) = \pi_*(X \wedge E)$. The coefficient ring of E is $E_*(S^0) = \pi_*(E)$ with multiplication defined by

$$\pi_*(E) \otimes \pi_*(E) \longrightarrow \pi_*(E \wedge E) \xrightarrow{m_*} \pi_*(E).$$

$E_*(X)$ is a right $E_*(S^0)$-module under the map

$$\begin{aligned} E_*(X) \otimes E_*(S^0) &= \pi_*(X \wedge E) \otimes \pi_*(E) \\ &\longrightarrow \pi_*(X \wedge E \wedge E) \\ &\longrightarrow \pi_*(X \wedge E). \end{aligned}$$

(2.2.65) <u>Homology operations</u>. $E_*(E)$ is a <u>two-sided</u> $E_*(S^0)$-module, and the two actions differ by the canonical involution $c:E_*(E) = \pi_*(E \wedge E) \longrightarrow \pi_*(E \wedge E) = E_*(E)$ induced by interchanging factors. We shall now require that the $E_*(E)$ be flat, that is, that the functors $— \otimes_{E_*(S^0)} E_*(E)$ and $E_*(E) \otimes_{E_*(S^0)} —$ from right $R_*(S^0)$-modules to right $E_*(S^0)$-modules be exact. This requirement holds for at least $S^0, K(Z_p)$, BO, BU, MO, MU, and MSp, though <u>not</u> for $K(Z)$.

Then $E_*(E)$ becomes a <u>Hopf algebra</u> over $E_*(S^0)$. The <u>product map</u>

$$\phi:E_*(E) \otimes E_*(E) \longrightarrow E_*(E)$$

is the composite

$$\begin{aligned} \pi_*(E \wedge E) \otimes \pi_*(E \wedge E) &\longrightarrow \pi_*(E \wedge E \wedge E \wedge E) \\ &\xrightarrow{(\mathrm{id} \wedge \mathrm{switch} \wedge \mathrm{id})_*} \pi_*(E \wedge E \wedge E \wedge E) \\ &\xrightarrow{(m \wedge m)_*} \pi_*(E \wedge E). \end{aligned}$$

Also, ϕ induces a product over $E_*(S^0)$ (easy).

The coproduct (or diagonal map)

$$\psi = \psi_E : E_*(E) \longrightarrow E_*(E) \otimes E_*(E)$$

is a special case of the coaction map

$$\psi_X : E_*(X) \longrightarrow E_*(X) \otimes E_*(E).$$

The coaction map is defined as follows:

$$\pi_*(X \wedge E) = \pi_*(X \wedge S^0 \wedge E)$$

$$\xrightarrow{(id \wedge i \wedge id)_*} \pi_*(X \wedge E \wedge E)$$

$$\xleftarrow{\cong} \pi_*(X \wedge E) \otimes_{\pi_*(E)} \pi_*(E \wedge E)$$

(The latter map is an isomorphism because $\pi_*(E)$ is flat [Adams - 3, p. 68, Lemma 1]).

It is now easy to extend the theory of coaction of the dual of the mod - p Steenrod algebra on $H_*(\ ,Z_p)$ to a theory of homology "operations" for E_*.

(2.2.66) Cohomology and cohomology operations are defined dually. $E^*(X)$ becomes a module over $E_*(S^0)$ as follows:

$$E^p(X) \otimes E^q(S^0) = [X \wedge S^{-p}, E] \otimes [S^q, E]$$

$$\longrightarrow [X \wedge S^{-p} \wedge S^q, E \wedge E]$$

$$\longrightarrow [X \wedge S^{-p+q}, E]$$

$$= E^{p-q}(X).$$

$E^*(E)$ is also a Hopf algebra, with the product

$$E^*(E) \otimes E^*(E) \longrightarrow E^*(E)$$

induced by the composition

$$\begin{aligned}[E \wedge S^p, E] \otimes [E \wedge S^q, E] &\\ &\longrightarrow [E \wedge S^p \wedge S^q, E \wedge S^q] \otimes [E \wedge S^q, E]\\ &\longrightarrow [E \wedge S^p \wedge S^q, E]\\ &= [E \wedge S^{p+q}, E].\end{aligned}$$

The (right) action of $E^*(E)$ on $E^*(X)$ is defined by a similar composition.

The power of the above very general theory is easily demonstrated in the following simple proof (due to Adams [Adams - 4]) that if $MU_*(X) = 0$, then $BU^*(X) = 0$. Recall that BU is a module spectrum over MU. Each map $X \wedge S^p \longrightarrow BU$ may be factored as follows:

$$\text{(2.2.67)} \qquad \begin{aligned} X \wedge S^p &\longrightarrow X \wedge S^p \wedge S^0\\ &\longrightarrow X \wedge S^p \wedge MU\\ &\longrightarrow BU \wedge MU\\ &\longrightarrow BU\,.\end{aligned}$$

But $\pi_*(X \wedge S^p \wedge MU) \cong \pi_*(X \wedge MU \wedge S^p) \cong \pi_{*-p}(X \wedge MU) = MU_{*-p}(X) = 0$. Thus $X \wedge S^p \wedge MU \simeq *$ by the Whitehead Theorem for CW spectra (see e.g., [Adams - 1]). Hence, the composite map (2.2.67) is null-homotopic, as required.

§2.3 Model categories.

We shall describe the basic properties of closed model categories and their associated homotopy categories; this theory is due to D. G. Quillen [Q -1, §§I.1 - I.5].

(2.3.1) Definition. An ordered pair of maps (i,p) is said to have the lifting property if given any solid-arrow commutative diagram

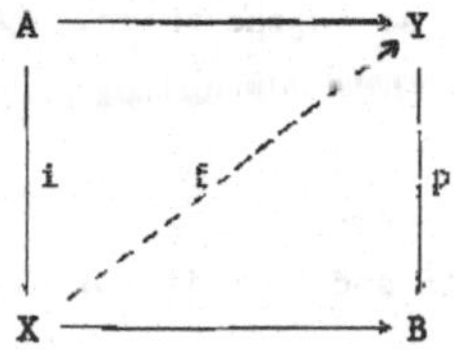

there exists a filler f.

(2.3.2) Definition. A closed model category consists of a category C together with three classes of maps in C, called the fibrations, cofibrations, and weak equivalences satisfying the following axioms.

M0. C is closed under finite colimits and limits.

M1. If a map i is a cofibration, a map p is a fibration, and either i or p is a weak equivalence, then the pair (i,p) has the lifting property.

M2. Any map f may be factored as f = pi where i is a cofibration and a weak equivalence and p is a fibration, or i is a cofibration and p is a fibration and a weak equivalence.

M3. Fibrations (resp. cofibrations) are stable under composition and base change (pullbacks) (resp., cobase change (pushouts)). Any isomorphism is a fibration and a cofibration.

M4. The base extension (resp., cobase extension) of a map which is both a fibration (resp., cofibration) and a weak equivalence is a weak equivalence.

M5. Let

$$X \xrightarrow{f} Y \xrightarrow{g} Z$$

be a diagram in C. If any two of the maps f,g , and gf are weak equivalences, then so is the third. Any isomorphism is a weak equivalence.

M6a. A map p is a fibration if and only if for all maps i which are cofibrations and weak equivalences, the pair (i,p) has the lifting property.

M6b. A map i is a cofibration if and only if for all maps p which are fibrations and weak equivalences, the pair (i,p) has the lifting property.

M6c. A map f is a weak equivalence if and only if f = uv where for all cofibrations i and fibrations p, the pairs (i,u) and (v,p) have the lifting property.

Observe that Axioms M5 and M6 imply Axioms M1, M3, and M4; hence to show that a given category is a closed model category it suffices to verify Axioms M0, M2, M5 and M6. We shall want the following technical definitions.

(2.3.4) Definitions. A map which is both a fibration (resp., cofibration) and a weak equivalence is called a trivial fibration (resp., trivial cofibration). The initial object of C shall be denoted ϕ ; the terminal object * (these objects exist by Axiom M0). An object X in C is called fibrant if the natural map $X \longrightarrow *$ is a fibration; cofibrant if the natural map $\phi \longrightarrow X$ is a cofibration.

(2.3.4) <u>Definition</u>. Let $X \in C$. A <u>cylinder object</u> for X consists of an object $X \otimes [0,1]$ and a commutative diagram

$$X \coprod X \xrightarrow{i_0 + i_1} X \otimes [0,1] \xrightarrow{p} X ,$$

where the map (i_0, i_1) is a cofibration, the map p is a weak equivalence, and $pi_0 = pi_1 = id_X$. We shall frequently write $X \otimes 0$ for $i_0(X)$ and $X \otimes 1$ for $i_1(X)$. <u>Caution</u>: in general, $X \otimes [0,1]$ is <u>not</u> the "tensor" product of X with an object $[0,1]$; in fact, in general, $X \otimes [0,1]$ need <u>not</u> depend functorially upon X.

We may use cylinder objects to form mapping cylinders with the usual properties. For example, we have the following.

(2.3.5) <u>Proposition</u>. Let $f: X \longrightarrow Y$ be a cofibration. Then there exist suitable cylinder objects so that f induces a trivial cofibration

$$Y \times 0 \cup X \otimes [0,1] \longrightarrow Y \otimes [0,1],$$

and a cofibration

$$Y \otimes 0 \cup X \otimes [0,1] \cup Y \otimes 1 \longrightarrow Y \otimes [0,1]. \qquad (2.3.6)$$

If f is a trivial cofibration, the induced map (2.3.6) is also a trivial cofibration.

<u>Proof</u>. Consider the commutative diagram

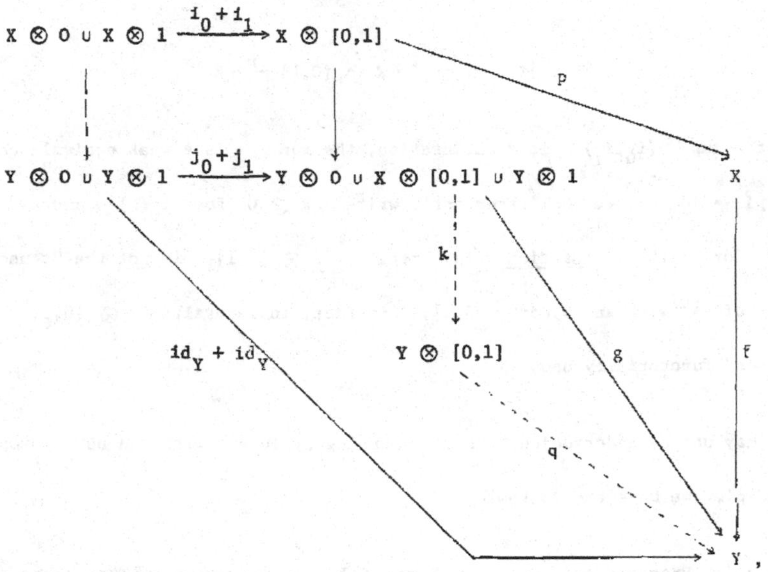

in which the subdiagram

$$X \otimes 0 \cup X \otimes L \xrightarrow{i_0 + i_1} X \otimes [0,1] \xrightarrow{p} X$$

is a cylinder object for X, $Y \otimes 0 \cup X \otimes [0,1] \cup Y \otimes 1$ is the pushout of the upper left square, and g is the map induced by the maps fp and $id_Y + id_Y$. The map $j_0 + j_1$ is the pushout (cobase extension) of the cofibration $i_0 + i_1$, hence is itself a cofibration. Factor g as qk where k is a cofibration and q is a trivial cofibration (dotted arrows above). We obtain a suitable cylinder object for Y, namely

$$Y \otimes 0 \cup Y \otimes 1 \xrightarrow{kj_0 + kj_1} Y \otimes [0,1] \xrightarrow{q} Y,$$

so that f induces cofibrations

$$Y \otimes 0 \cup X \otimes [0,1] \longrightarrow Y \otimes [0,1],$$

$$Y \otimes 0 \cup X \otimes [0,1] \cup Y \otimes 1 \longrightarrow Y \otimes [0,1].$$

The remaining assertions are easily checked by applying Axiom M5; details are omitted. □

We shall discuss cocylinder objects (dual to cylinder objects) in C, and loop and suspension functors as well as the induced cofibration and fibration sequences in §3.4 below.

In these notes we shall always assume that our closed model categories C satisfy the following niceness condition.

<u>Condition N</u>:

<u>N1</u>. Each cofibration is a pushout of a cofibration of cofibrant objects.

<u>N2</u>. Each fibration is a pullback of a fibration of fibrant objects.

<u>N3</u>. At least one of the following statements hold:

<u>N3a</u>. All objects are cofibrant.

<u>N3b</u>. All objects are fibrant.

<u>N4</u>. There exist functorial cylinder objects, denoted by $-\otimes[0,1]$ with $i_0(-) = -\otimes 0$ and $i_1(-) = -\otimes 1$.

The following closed model categories satisfy Condition N.

SS; (D. M. Kan [Kan -3], [Kan -4], see D. Quillen [Q- 1, §II.3]). Since all objects are cofibrant, Condition N(1) is trivial. N(2) is due to J. C. Moore, see [Q -1, §II.3]. For N(3), let $X \otimes [0,1] = X \times \Delta^1$, the usual product.

Top; the category of topological spaces with the following structure: cofibrations and fibrations are defined by the homotopy-extension and covering-homotopy properties, respectively; weak equivalences are ordinary homotopy equivalences. This is due to A. Strøm [Str]. Condition N is clear.

CG; the category of compactly generated spaces, with a similar structure. See N. E. Steenrod [St - 3]; also, [Has - 3].

Sing; the category of topological spaces with the following <u>singular</u> structure: cofibrations are pushouts of inclusions of subcomplexes of CW complexes, fibrations are Serre fibrations, weak equivalences are weak homotopy equivalences [Q - 1, §II.3]. Again, Condition N is clear.

SSG (SSAG); Simplicial groups (resp., simplicial abelian groups) [Q -1, §II.3]. Here let

$$X \times [0,1] = F(X \times [0,1])/F(X \times 0) \sim X \times 0,\ F(X \times 1) \sim X \times 1,$$

where the products are taken in SS and F is the free (resp., free abelian) simplicial group functor. Condition N follows from Condition N for SS.

Sp; D. M. Kan's simplicial spectra [Kan -1]. K. Brown [Brown] first proved that Sp is a closed model category with a closed model structure similar to that on SS. See §2.2. Condition N follows from condition N for SS.

We shall now describe the homotopy theory of a closed model category C which satisfies condition N.

(2.3.7) <u>Definition</u> [Q - 1]. The homotopy category of C, Ho(C), is the quotient category obtained from C by inverting all weak equivalences.

Quillen proved the following.

(2.3.8) <u>Proposition</u> [Q-1, Prop. I.5.1]. A map f in C becomes an equivalence in $Ho(C)$ if and only if f is a weak equivalence in C.

The following homotopy theory is required for the proof.

(2.3.9) <u>Definition</u>. If X is cofibrant and Y is fibrant, maps $f,g:X \rightrightarrows Y$ will be called <u>homotopic</u> (denoted $f \simeq g$) if there is a map $H:X \times [0,1] \longrightarrow Y$ with $H|_{X\times 0} = f$ and $H|_{X\times 1} = g$. Compare [Q-1, §I.1].

(2.3.10) <u>Definition</u>. Let C_{cf} denote the full subcategory of cofibrant, fibrant objects in C.

(2.3.11) <u>Proposition</u> [Q-1, Lemma 1.5.1 and its dual]. A map $f:X \longrightarrow Y$ in C_{cf} is a weak equivalence if and only if there is a map $g:Y \longrightarrow X$ with $gf \simeq id_X$ and $fg \simeq 1_Y$.

The proof is analogous to that of [Q-1, Lemma I.5.1], and is omitted. □

<u>Proof of Proposition (2.3.8)</u>. Given a map $f:X \longrightarrow Y$ in C which is invertible in $Ho(C)$, form a commutative diagram

$$\begin{array}{ccc} X & \xrightarrow{\ f\ } & Y \\ \uparrow & & \uparrow \\ X' & & Y' \\ \downarrow & & \downarrow \\ X'' & \overset{[f'']}{\dashrightarrow} & Y'' \end{array},$$

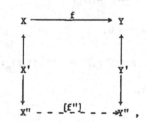

where X' and Y' are cofibrant, X'' and Y'' are both cofibrant and fibrant, and all vertical maps are weak equivalence. Then use the axioms to realize the map [f"] in Ho(C) by a map f" in C. Finally, Proposition (2.3.11) implies that f" is a weak equivalence. See [Q -1, §II.5] for details.

§2.4. Simplicial closed model categories.

We shall discuss function spaces in SS, and their generalization to the concept of simplicial closed model category given by Quillen [Q - 1, §§II.2.2].

The product in SS is given by $X \times Y = \{(X \times Y)_n = X_n \times Y_n\}$, together with the induced face and degeneracy maps. This product is coadjoint to the "function space" (or internal mapping functor) $HOM(X,Y) = \{HOM(X,Y)_n = SS(X \times \Delta^n,Y)\}$, together with the face and degeneracy maps induced form the maps $d^i:\Delta^{n-1} \longrightarrow \Delta^n$ and $s^i:\Delta^{n+1} \longrightarrow \Delta^n$, $0 \leq i \leq n$. See e.g., [May -1] or [Q -1]. In the context of the adjoint pair $(\times, HOM)$ we shall write $\otimes$ for $\times$.

The functors $\otimes$ and HOM satisfy the following properties.

(i) There is an associative composition

$HOM(X,Y) \times HOM(Y,Z) \longrightarrow HOM(X,Z)$, for all

X, Y, and Z in C, and a natural isomorphism of

functors $SS(X,Y) \xrightarrow{\cong} HOM(X,Y)_0$,

$$u \longmapsto \tilde{u},$$

such that for u in SS(X,Y), f in $HOM(Y,Z)_n$,

and g in HOM (W,X),

$$f \circ (S_0)^n \bar{u} = \mathrm{HOM}\ (u,Z)_n(f), \quad \text{and}$$

$$(S_0)^n \bar{u} \circ g = \mathrm{HOM}\ (W,u)_n(g)$$

$$SS(X,Y) \xrightarrow{\cong} \mathrm{HOM}\ (X,Y)_0,$$

for all Y in C .

(ii) There are natural maps $\alpha : Y \longrightarrow \mathrm{HOM}\ (X, X \otimes Y)$ which induce isomorphisms (enriched adjunction [E - K])

$$\mathrm{HOM}\ (X \otimes Y, Z) \xrightarrow{\cong} \mathrm{HOM}\ (Y, \mathrm{HOM}\ (X,Z)),$$

for all X and Z in C .

(iii) For all Y in C there are natural maps

$\beta : Y \longrightarrow \mathrm{HOM}\ (\mathrm{HOM}\ (Y,Z), Z)$ which induce isomorphisms

$$\mathrm{HOM}\ (X, \mathrm{HOM}\ (Y,Z)) \xrightarrow{\cong} \mathrm{HOM}\ (Y, \mathrm{HOM}\ (X,Z)),$$

for all X and Z in C .

(iv) There are natural isomorphisms

$$\mathrm{HOM}\ (*, X) \xrightarrow{\cong} X,$$

for all X in C.

Consequently,

(v) The composition maps in (i) and SS are compatible; i.e., the following diagram commutes for all X,Y,Z in SS :

$$\begin{array}{ccc} \mathrm{HOM}\,(X,Y)_0 \times \mathrm{HOM}\,(Y,Z)_0 & \longrightarrow & \mathrm{HOM}\,(X,Z)_0 \\ \Big\downarrow \cong & & \Big\downarrow \cong \\ SS(X,Y) \times SS(Y,Z) & \longrightarrow & SS(X,Z)\ . \end{array}$$

(vi) $\otimes$ is coherently associative and commutative, with $*$ as coherent unit.

Function spaces and products in CG have similar properties. For another example, define $\otimes$ on SS_* by $X \otimes Y = X \wedge Y = X \times Y/X \vee Y$, and HOM as above except that Δ^n is replaced by Δ^{n*} obtained from Δ^n by adjoining a disjoint basepoint. These ideas have been abstracted in S. Eilenberg and G. M. Kelly's closed symmetric monoidal category [E-K].

The singular closed model structure on Top, Top_{sing} ([Q-1, §II.3], see §2.3) admits a "singular function space." For X and Y in Top, define HOM (X,Y) to be the simplicial set with $\mathrm{HOM}\,(X,Y)_n = \mathrm{Top}\,(X \times R\Delta^n, Y)$ together with the induced face and degeneracy maps, see above. Here R denotes Milnor's geometric realization functor [Mil-2], see [May-1]. For X in Top and K in SS, define $X \otimes K = X \times RK$. Then

$$\mathrm{HOM}\,(X \otimes K, Y) = \mathrm{HOM}\,(K, \mathrm{HOM}\,(X,Y))$$

for K in SS, and X,Y in Top. Quillen generalized this concept by introducing closed simplicial model categories, described below.

(2.4.1) Definition (see [Q-1, Definition §II.1.1]) A simplicial category is a category C together with the following structure:

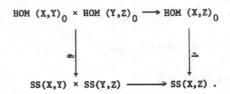

(i) A functor HOM $(-,-)$ from $C \times C$ to SS, contravariant in the first variable and covariant in the second.

(ii) For an X, Y, and Z in C, maps in SS

$$\mathrm{HOM}\,(X,Y) \times \mathrm{HOM}\,(Y,Z) \longrightarrow \mathrm{HOM}\,(X,Z)$$

called composition.

(iii) An isomorphism of functors

$$C(X,Y) \xrightarrow{\cong} \mathrm{HOM}\,(X,Y)_0,$$

$$u \longrightarrow \tilde{u},$$

where $\mathrm{HOM}\,(X,Y)_0$ consists of the 0-simplices of HOM (X,Y).

These functors are required to satisfy the following conditions.

(1) Composition is associative.

(2) For u in C(X,Y), f in $\mathrm{HOM}\,(Y,Z)_n$, and g in $\mathrm{HOM}\,(W,X)_n$,

$$f \circ (S_0)^n \tilde{u} = \mathrm{HOM}\,(u,Z)_n(f), \quad \text{and}$$

$$(S_0)^n \tilde{u} \circ g = \mathrm{HOM}\,(W,u)_n(g).$$

(2.4.2) <u>Definition</u> (see [Q-1, Definition §II.1.3]). For X in C and K in SS, $X \otimes K$ shall denote an object of C together with a distinguished map $\alpha: K \longrightarrow \mathrm{HOM}\,(X, X \otimes K)$ which induces a natural isomorphism

$$\mathrm{HOM}\,(X \otimes K, Y) \xrightarrow{\cong} \mathrm{HOM}\,(K, \mathrm{HOM}\,(X,Y)).$$

HOM (K,X) shall denote an object of C together with a distinguished map $\beta: K \longrightarrow \mathrm{HOM}\,(\mathrm{HOM}\,(K,X), X)$ which induces a natural isomorphism

$$\text{HOM }(Y, \text{HOM }(K,X)) \xrightarrow{\cong} \text{HOM }(K, \text{HOM }(Y,X)).$$

(2.4.3) Examples. Clearly SS, with its usual symmetric monoidal structure, is a simplicial category (see [Q -1], [May], [E -K]). Top_{sing} (see §2.3) is also a simplicial category.

(2.4.4) Definition [Q- 1, Definition §II.2.2]. A closed simplicial model category consists of a closed model category C which is also a simplicial category satisfying the following two conditions.

SM0. For X in C and K a *finite* simplicial set, then $X \otimes K$ and HOM (K,X) exist.

SM7. If $i:A \to X$ is a cofibration in C and $p:Y \to B$ is a fibration in C, then

$$\text{HOM }(X,Y) \longrightarrow \text{HOM }(A,Y) \times_{\text{HOM }(A,B)} \text{HOM }(X,B)$$

is a fibration in SS which is trivial if either i or p is trivial.

Recall that for spaces X and Y, say in CG,

$$[X,Y] \cong \pi_0(\text{HOM }(X,Y)),$$

where HOM (X,Y) is the usual function space. Similarly, for X and Y in SS with Y fibrant (i.e., Kan), $[X,Y] \cong \pi_0(\text{HOM }(X,Y))$. A similar statement holds in an abstract simplicial closed model category C.

(2.4.4) Proposition [Q -1, Proposition §II.2.5]. If X is cofibrant in C and Y is fibrant in C, then

$$Ho(C)(X,Y) \cong \pi_0(HOM\ (X,Y)),$$

the set of path components of HOM (X,Y). □

(2.4.6) Remarks.

a) Our use of HOM in three settings, on $C \times C$, $C \times S$, and $S \times S$, should emphasize the analogy between the functor HOM of Definition and the usual "function space" (internal mapping) functors.

b) All of the above results have pointed analogues; replace SS by SS_* and Δ^n by Δ^{n*}, which is obtained from Δ^n by adjoining a disjoint basepoint.

c) In general the function space in Top or CG, HOM (X,Y), is not homotopy equivalent to the realization of the singular "function space" $R(HOM_{sing}(X,Y)) \equiv R\{Top\ (X \times R\Delta^n, Y),\ d_i, s_i\}$. For example, the latter space is always a CW complex.

§2.5. Homotopy theories of pro-spaces.

In this section we shall briefly indicate the need for a "sophisticated" homotopy theory of pro-spaces. M. Artin and B. Mazur took pro-Ho(Top) to be the homotopy theory of pro-Top. Unfortunately, this point of view is inadequate for some purposes. For example, Quillen [Q-1, C. II, p. 0.3] observed that the category pro-Ho(Top) was not the homotopy category of a model structure on pro-Top. One next attempts to define homotopy globally in pro-Top, that is, to call maps $f,g:\{X_i\} \longrightarrow \{Y_j\}$ homotopic if there is a homotopy

$H:\{X_i\} \times [0,1] \equiv \{X_i \times [0,1]\} \longrightarrow \{Y_j\}$ from f to g. This notion is stronger than the Artin-Mazur notion which would identify two level maps $\{X_n \xrightarrow{f_n, g_n} Y_n\}$ if there were homotopies $H_n : f_n \simeq g_n$ without any coherence criteria among the H_n. For example, let D denote the inverse system

$$S^1 \xleftarrow{2} S^1 \longleftarrow \cdots,$$

where $S^1 \xleftarrow{2} S^1$ is the degree two map $z \longmapsto z^2$. Then, there is a unique Artin-Mazur homotopy class of maps from a point to D, but there are uncountably many global homotopy classes of maps from a point to D (more precisely, $\lim^1 \{Z \xleftarrow{2} Z \xleftarrow{2} \cdots\}$ such classes). The following example shows that the notion of global homotopy is also too naive.

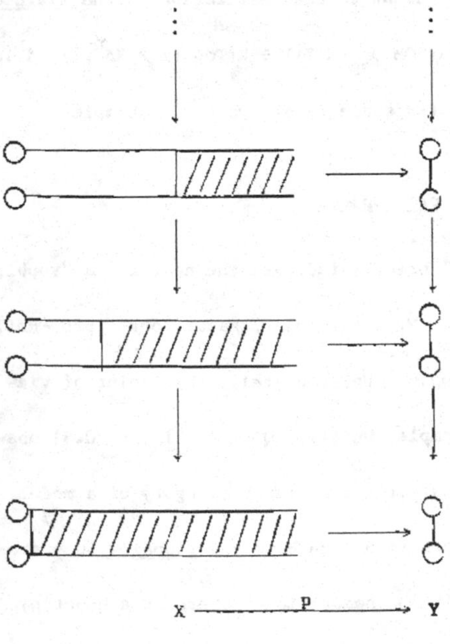

;

here $X = \{X_n \equiv (S^1 \vee [0,\infty)) \times \{0,1\} \cup [n,\infty) \times [0,1]\}$, and $Y = \{Y_n \equiv S^1 \times \{0,1\} \cup [0,1]\}$. The map $p = \{p_n\}:X \longrightarrow Y$ is levelly a homotopy equivalence, but there is no homotopy inverse to p in pro-Top. If, however, the bonding maps of the towers X and Y are fibrations, then the notion of global homotopy turns out to be the "right" notion. The "right" homotopy category, Ho(pro-Top), is defined by formally inverting level homotopy equivalences. We shall define Ho(pro-Top) in §3.

§3. THE MODEL STRUCTURE ON PRO-SPACES

§3.1. Introduction.

In this chapter we shall associate to a closed model category C which satisfies condition N (§2.3) a natural closed model structure on pro-C. This chapter is organized as follows.

In §3.2, we discuss the homotopy theory of C^J, where J is a cofinite strongly directed set ($a \leq b$ and $b \leq a \Rightarrow a = b$). We shall develop a closed model structure on C^J (Theorem (3.2.2)) which is natural in the following sense (Theorem (3.2.4)). The constant diagram functor $C \to C^J$ preserves the model structure. The inverse limit functor $\lim: C^J \longrightarrow C$ preserves fibrations and trivial fibrations.

In §3.3, we shall extend the closed model structure from the level categories C^J to pro-C (Theorem (3.3.3)) with the same naturality properties as our closed model structure on C^J (Theorem (3.3.4)). We also obtain a natural closed model structure on the full subcategory tow-C ⊂ pro-C. Simplicial structures on pro-C are discussed in §3.5.

§3.4 is concerned with suspension and loop functors, and cofibration and fibration sequences. D. Quillen [Q-1, §§1.2-3] developed a general theory of suspension and loop functors, and cofibration and fibration sequences in the homotopy category of an abstract closed model category. We shall sketch this theory in the context of Ho(pro-C). We shall show that an inverse system of fibrations over

C is equivalent in Ho(pro - C) to a short fibration sequence.

In §3.6 we consider the category Maps (pro - C) and a full subcategory (C, pro - C) whose objects are maps $A \longrightarrow X$ (in pro - C) with X stable in pro - C.

We develop useful (see §§6-8) geometric models of Ho(Top, tow - Top) and Ho(tow - Top) in §3.7.

We shall compare our closed model structure to those of A. K. Bousfield and D. M. Kan [B - K, p. 314] and J. Grossman [Gros - 1] in Remarks (3.2.5).

The above theory of pro - spaces admits an evident dualization to direct systems (inj - spaces). We shall briefly sketch this theory in §3.8.

§3.2. The homotopy theory of C^J.

Let C be a closed model category which satisfies Condition N (§2.3). Let $J(=\{j\})$ be a cofinite strongly directed set. We shall show that C^J inherits a natural closed model structure from C; this will yield the required homotopy category $Ho(C^J)$ (see §2.3).

(3.2.1) Definitions. A map $f:X \longrightarrow Y$ $(=\{f_j:X_j \longrightarrow Y_j\})$ in C^J is a cofibration (resp., weak equivalence) if for all j in J, the maps f_j are cofibrations (resp., weak equivalences) in C.

A map f in C^J is a fibration if it has the right-lifting-property with respect to all maps i which are both cofibrations and weak equivalences.

The main result of this section is the following.

(3.2.2) <u>Theorem</u>. C^J, together with the above structure, is a closed model category.

A map $f: X \to Y$ in C^J is a fibration if for each j in J, the induced map q_j in the diagram

(3.2.3)

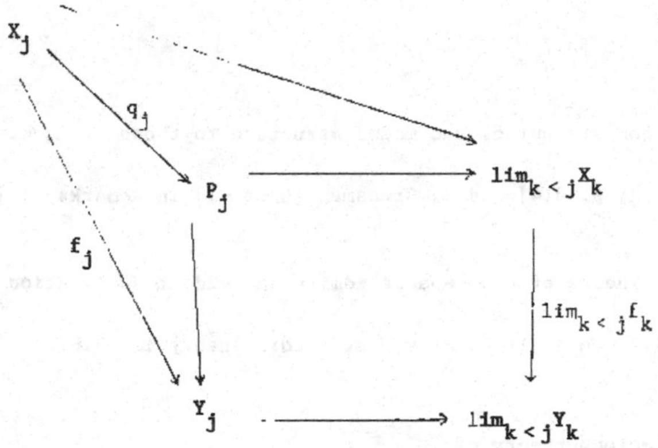

(P_j is the pullback) is a fibration.

The special case $J = N$ is used to obtain a closed model structure in tow - C in §3.3.

(3.2.4) <u>Theorem</u>. The constant diagram functor $C \to C^J$ preserves cofibrations, fibrations, and weak equivalences. The inverse limit functor $\lim: C^J \to C$ preserves fibrations, and trivial fibrations (maps which are both fibrations and weak equivalences).

<u>Proof</u>. Immediate from Definitions (3.2.1). □

(3.2.5) Remarks. Bousfield and Kan [B - K, p. 314] defined a different closed model structure on C^J by defining fibrations and weak equivalences degreewise, and defining cofibrations by the appropriate lifting property. The Bousfield-Kan structure has the disadvantage that most of Theorem (3.2.4) is false: only fibrations and weak equivalences are preserved by the constant diagram functor; none of the model structure is preserved by the inverse limit functor. Consequently, our homotopy inverse limit functor (§4.2) is simpler than theirs; this simplicity makes evident the applications to homological algebra in §§4.5-4.8, see also below. The Bousfield-Kan structure is natural on direct systems, see §3.8, especially, (3.8.1)-(3.8.4).

J. Grossman [Gros -1] also introduced a closed model structure on the category of towers of simplicial sets. His structure is weaker than ours; essentially, he inverts $\natural$-isomorphisms in the sense of [A - M]. See (5.4.4)-(5.4.5) for the definition of $\natural$-isomorphism.

Our definition of fibration was motivated by the definition of a cofibration of pairs (for the inclusion $(X,A) \longrightarrow (Y,B)$ to be a cofibration one usually asks that the induced map $X \cup_A B \longrightarrow Y$ be a cofibration), and the analogous definition of a cofibration of CW spectra (see [Vogt -2]). Also:

a) Our definition of fibration is consistent with the definition of a flasque pro - group (see §§4.4, 4.8).

b) The associated definition of cofibration means that a proper cofibration $X \longrightarrow Y$ induces a cofibration of the ends $\varepsilon(X) \longrightarrow \varepsilon(Y)$ (see §6).

The proof of Theorem (3.2.2) is contained in Propositions (3.2.6), (3.2.24), (3.2.27) and (3.2.28), below.

(3.2.6) Proposition. (Verification of Axiom M0). C^J admits finite limits and colimits.

Proof. Let D be a finite diagram in C^J. The induced diagrams D_J over C have colimits $\operatorname{colim} D_j$ and limits $\lim D_j$ in C by Axiom M0 for C. These yield objects $\{\operatorname{colim} D_j\}$ and $\{\lim D_j\}$ in C^J which are easily seen to be the colimit and limit of D, respectively. □

In fact, if C admits more general colimits or limits, so does C^J.

In order to verify Axioms M2, M5 and M6 for C^J we shall give explicit descriptions of fibrations (Proposition (3.2.7)) and trivial fibrations (Proposition (3.2.17)) in C^J. Our descriptions will involve diagram (3.2.3).

(3.2.7) Proposition. A map $p:Y \to B$ in C^J is a fibration if and only if for each j in J the induced map q_j in the diagram

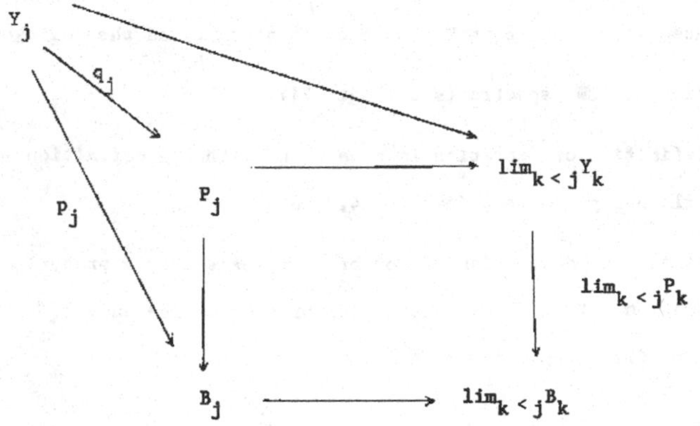

(P_j is the pullback) is a fibration in C.

<u>Proof</u>. First, let p be a fibration in C^J, that is, assume that p has the right-lifting-property with respect to the class of trivial cofibrations in C^J. We shall show that each induced map $q_j : Y_j \longrightarrow P_j$ has the same right-lifting-property by constructing suitable "test maps" $K \longrightarrow L$ in C^J which are trivial cofibrations.

Consider a solid-arrow commutative diagram

(3.2.8)
$$\begin{array}{ccc} A & \longrightarrow & Y_j \\ {\scriptstyle i}\downarrow & \overset{f}{\nearrow} & \downarrow{\scriptstyle q_j} \\ X & \longrightarrow & P_j \end{array}$$

in C in which i is a trivial cofibration. Define objects $K = K_k$ and $L = L_k$ in C^J as follows:

$$K_k = \begin{cases} X & \text{for } k < j, \\ A & \text{for } k = j, \\ \phi & \text{otherwise}; \end{cases}$$

$$L_k = \begin{cases} X & \text{for } k \leq j, \\ \phi & \text{otherwise}. \end{cases}$$

The required bonding maps are induced by i and id_X. Then there is an induced trivial cofibration $i' : K \longrightarrow L$ in C^J. Diagram (3.2.8) induces a solid-arrow commutative diagram

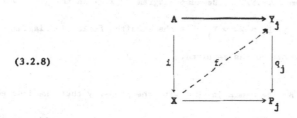

(3.2.9)

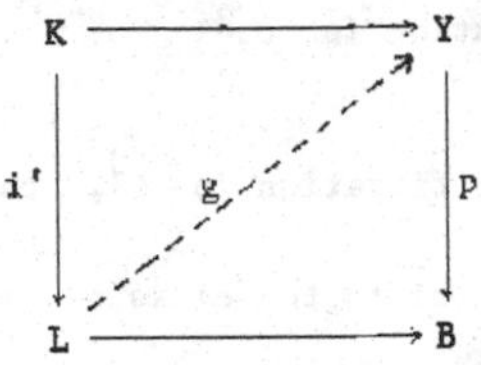

in C^J (the maps $K_k \longrightarrow Y_k$ are induced from the map $A \longrightarrow Y_j$ for $j = k$ and the composite maps $X \longrightarrow P_j \longrightarrow Y_k$ for $k < j$ (see diagram (3.2.8)), the other maps in diagram (3.2.9) are defined similarly). The right-lifting-property of p yields a filler g in diagram (3.2.9). Because diagram (3.2.8) is the jth level of diagram (3.2.9), the map $g_j : L_j = X \longrightarrow Y_j$ is the required filler in diagram (3.2.8). Hence q_j is a fibration, as required.

Conversely, let $p : Y \longrightarrow B$ be a map in C^J with the property that the induced maps $q_j : Y_j \longrightarrow P_j$ (see diagram (3.2.3)) are fibrations. Consider a solid-arrow commutative diagram in C^J

(3.2.10)

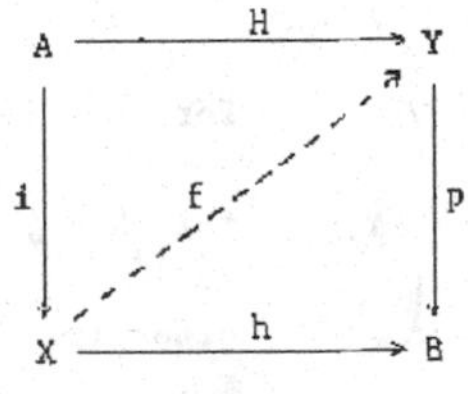

in which the map i is a trivial cofibration. We shall obtain the required filler

$$f = \{f_j : X_j \longrightarrow Y_j\}$$

by induction on j.

Consider a fixed index j. Suppose that for all $k < j$, there exist maps $f_k : X_k \longrightarrow Y_k$ with the following properties:

$$f_k i_k = H_k\,,$$

(3.2.11)

$$p_k f_k = h_k\,;$$

(3.2.12) $f_\ell \circ \text{bond} = \text{bond} \circ f_k$ for $\ell < k$ (if j has no predecessors these restrictions are vacuous). Formula (3.2.12) yields a composite map

(3.2.13) $$X_j \longrightarrow \lim_{k<j} Y_k \longrightarrow \lim_{k<j} B_j\,;$$

by formulas (3.2.11) this map is equal to the composite map

(3.2.14) $$X_j \longrightarrow B_j \longrightarrow \lim_{k<j} B_k.$$

In fact, formulas (3.2.10) - (3.2.14) yield a solid-arrow commutative diagram

(3.2.15) $$\begin{array}{ccc} A_j & \xrightarrow{H_j} & Y_j \\ {\scriptstyle i_j}\downarrow & \overset{f_j}{\nearrow} & \downarrow{\scriptstyle q_j} \\ X_j & \longrightarrow & P_j \end{array}$$

in C. (This uses the definition of P_j in diagram (3.2.3).) Because i_j is a trivial cofibration and g_j is a fibration in C, there exists a filler f_j in diagram (3.2.15). Further, diagram (3.2.15) yields formulas (3.2.11) and (3.2.12) for f_j.

By continuing inductively, we obtain the required filler $f = \{f_j\}$ in diagram (3.2.10). □

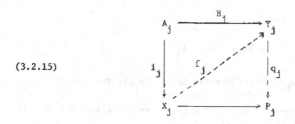

(3.2.16) <u>Proposition</u>. Let $p:X \longrightarrow B$ be a fibration in C^J. Then at each level, the map $p_j:Y_j \longrightarrow B_j$ is a fibration in C.

<u>Proof</u>. For a given j in J, consider the commutative diagram

(3.2.3)

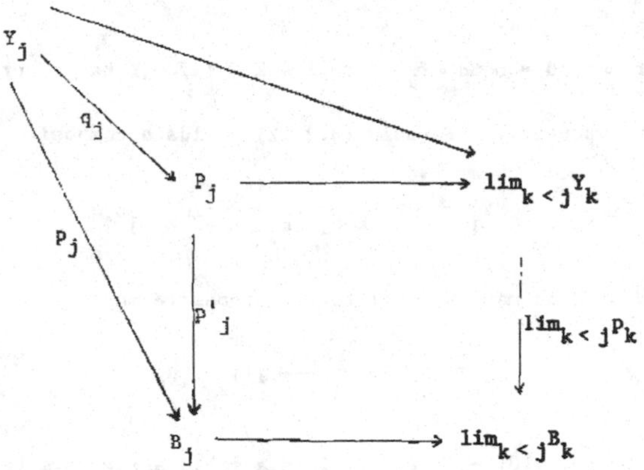

By introducing the indexing category

$$K = \{k \mid k < j\} \subset J,$$

we see that the map $\lim_{k<j} p_k$ is a fibration in C (apply Theorem (3.2.4)). Hence p'_j, and thus the composite map $p_j = p'_j q_j$ are fibrations in C. □

(3.2.17) <u>Remarks</u>. The above proof illustrates the usefulness of <u>cofinite strongly directed</u> indexing sets: maps in C^J may be constructed inductively.

Trivial fibrations in C^J admit a similar characterization.

(3.2.18) <u>Proposition</u>. A map $p:Y \longrightarrow B$ in C^J is a trivial fibration if and only if the induced maps $q_j:Y_j \longrightarrow P_j$ (see Proposition (3.2.6)) are trivial fibrations in C.

<u>Proof</u>. First, let $p:Y \longrightarrow B$ be a trivial fibration. If j is an initial element in J, the maps $q_j:Y_j \longrightarrow P_j$ and $p_j:Y_j \longrightarrow B_j$ are equal. But p_j is a fibration by Proposition (3.2.7) and weak equivalence by hypothesis. Hence $q_j(= p_j)$ is a trivial fibration, as required.

Now suppose that for a fixed (non-initial) j in J, and for all $k < j$, the induced maps q_k are trivial fibrations. We shall show that q_j is a trivial fibration.

Consider the commutative diagram

$$\begin{array}{ccccc}
Y & \xrightarrow{q_j} & P_j & \longrightarrow & \lim_{k<j} Y_k \\
 & \searrow^{f_j} & \downarrow & & \downarrow{\scriptstyle \lim_{k<j} p_k} \\
 & & B_k & \longrightarrow & \lim_{k<j} B_k
\end{array}$$

in which P_j is a pullback. If $\lim_{k<j} p_k$ were a trivial fibration, the induced map $P_j \longrightarrow B_k$ would also be a trivial fibration (by Axiom M4 for C), and hence a weak equivalence. But then q_j would be a weak equivalence (since f_j is a weak equivalence by hypothesis, this follows from Axiom M5 for C). But f_j is a fibration by Proposition (3.2.7), so that f_j would be a trivial fibration, as required. It therefore suffices to show that the maps

(3.2.19) $$\lim_{k<j} p_k : \lim_{k<j} Y_k \longrightarrow \lim_{k<j} B_k$$

are trivial fibrations.

To do this, we consider a commutative solid-arrow diagram of the form

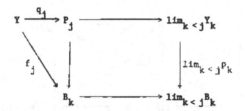

(3.2.20)

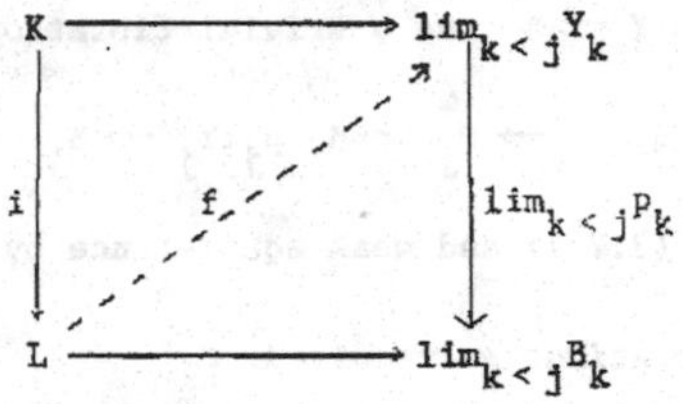

in C in which i is a cofibration. We may define a filler f in diagram (3.2.19) by defining maps

$$f_k : L \longrightarrow Y_k$$

for $k < j$ which make the diagrams

(3.2.21)

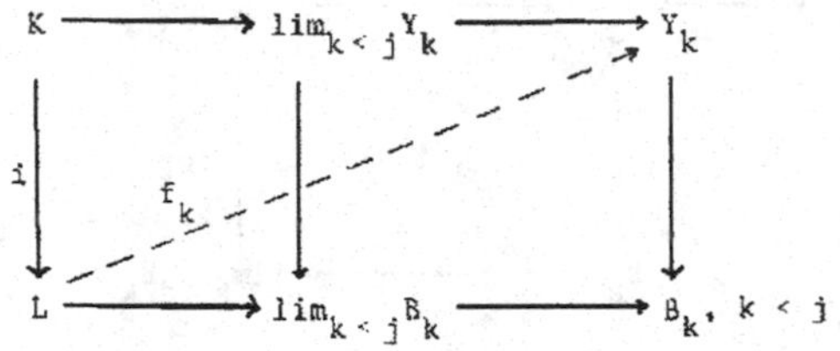

and

(3.2.22)

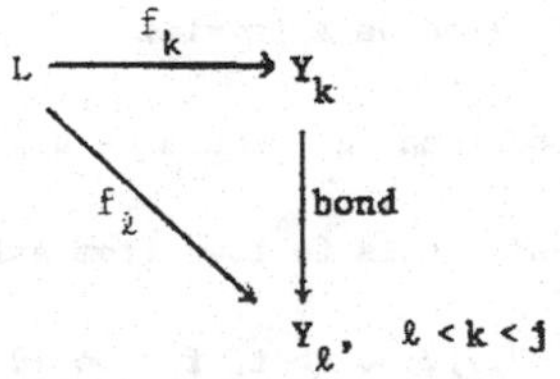

commute. This requires a second induction; this time on k. Suppose that for a fixed k, and all $\ell < k$ there exist the required fillers f_ℓ (if k has no predecessors, this condition is vacuous). We obtain a map

$$K \longrightarrow P_k$$

(see diagram (3.2.3)) for which the solid-arrow diagram

(3.2.23)

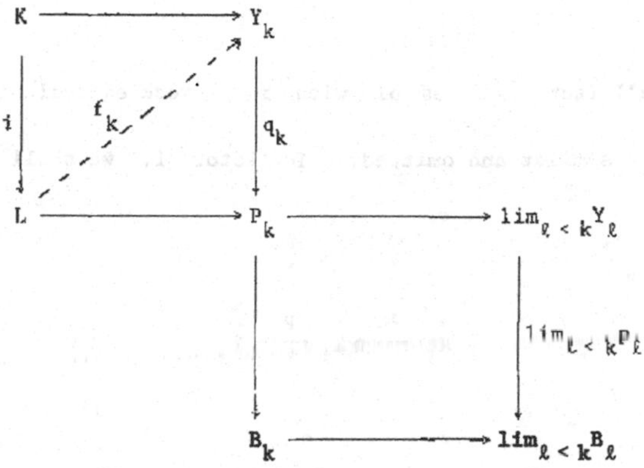

commutes. But the map q_k is a trivial fibration by our inductive assumption. Hence there exists a filler f_k for the upper left corner of diagram (3.2.23). Diagram (3.2.23) immediately implies that the required diagrams (3.2.21) and (3.2.22) commute. Continuing inductively yields the required maps f_k and filler

$$f = \lim_{k < j} f_k$$

in diagram (3.2.20). Hence, the maps (3.2.19) are trivial fibrations, as required.

The proof of the converse is similar to the proof of the "if" part of Proposition (3.2.6) and is omitted. □

We may now verify that C satisfies Axioms M2, M5, and M6 for a closed model category.

(3.2.24) <u>Proposition</u>. (Verification of Axiom M2) Any map $f: X \to Y$ in C^J may be factored as

$$X \xrightarrow{i} Z \xrightarrow{p} Y$$

where i is a cofibration, p is a fibration, and either i or p is a weak equivalence.

Proof. We shall factor f as pi with i a weak equivalence. The proof of the other case is similar and omitted. To factor f, we shall factor the maps $f_j:X_j \longrightarrow Y_j$ as

(3.2.25) $$X_j \xrightarrow{i_j} Z_j \xrightarrow{p_j} Y_j$$

where:

a) for $k < j$, i_j and p_j cover i_k and p_k, respectively;

b) the maps i_j are trivial cofibrations; and

c) the induced maps $q_j:Z_j \longrightarrow P_j$ associated with the maps $p_j:Z_j \longrightarrow Y_j$ (see diagram (3.2.3)) are fibrations.

Suppose for a given j and for all $k < j$, the maps f_k have been so factored. If j has no predecessors, this condition is vacuous. We may then form the commutative diagram

(3.2.26)

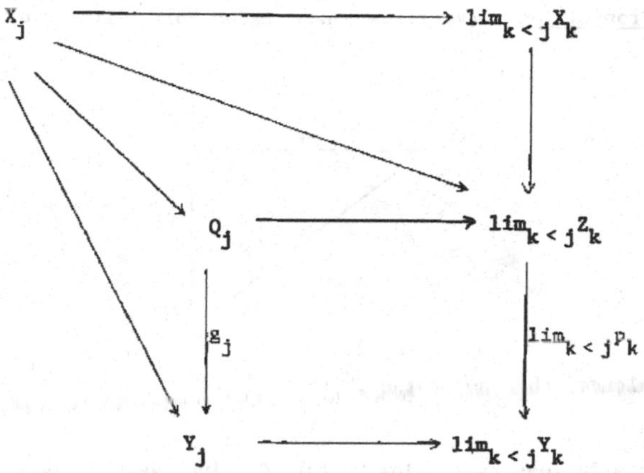

where Q_j is a pullback, and the map g_j is induced from the map $\lim_{k<j} p_k$. As in the proof of Proposition (3.2.17), see diagram (3.2.20), the map $\lim_{k<j} p_k$ is a fibration. Hence the induced map g_j (see diagram (3.2.26)) is a fibration (by Axiom M3 for C).

Now, factor the map $X_j \to Q_j$ (in diagram (3.2.26)) as the composite

$$X_j \xrightarrow{i_j} Z_j \xrightarrow{q'_j} Q_j$$

(using Axiom M2 for C). Finally, let

$$p_j = g_j q'_j : Z_j \longrightarrow Q_j \longrightarrow Y_j .$$

Then the factorization

$$X_j \xrightarrow{i_j} Z_j \xrightarrow{p_j} Y_j$$

(see diagram (3.2.25)) has the required properties. The conclusion follows. □

(3.2.27) <u>Proposition</u>. (Verification of Axiom M5). If two of the maps in the diagram

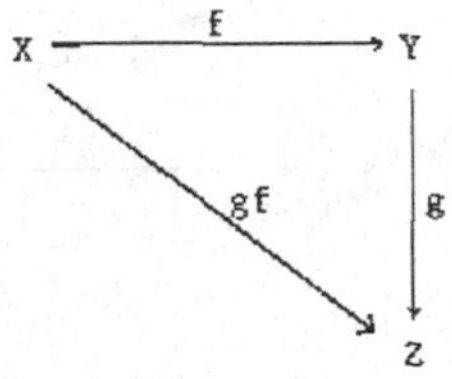

are weak equivalence, then so is the third.

<u>Proof</u>. This follows from Axiom M5 for C since weak equivalences in C^J are defined degreewise. □

(3.2.28) <u>Proposition</u>. (Verification of Axiom M6).

a) A map is a fibration if and only if it has the right-lifting-property with respect to the class of trivial cofibrations.

b) A map is a cofibration if and only if it has the left-lifting-property with respect to the class of trivial fibrations.

c) A map f is a weak equivalence if and only if $f = uv$ where v has the left-lifting-property with respect to the class of fibrations and u has the right-lifting-property with respect to the class of cofibrations.

<u>Proof</u>. Part a) is contained in Definitions (3.2.1).

The proof of Part b) is similar to that of Proposition (3.2.7). The main step is to associate to an element j of J and a fibration $Y \longrightarrow B$ in C, the objects $E = \{E_k\}$ and $A = \{A_k\}$ in C^J defined by setting

$$E_k = \begin{cases} Y & \text{for } k \geq j, \\ * & \text{otherwise,} \end{cases}$$

$$A_k = \begin{cases} B & \text{for } k \geq j, \\ * & \text{otherwise,} \end{cases}$$

with the evident bonding maps and also the fibration $E \longrightarrow B$ in C^J. One may then proceed as in the discussion following diagram (3.2.8). Remaining details are omitted.

For Part c), first let f be a weak equivalence. Factor f as uv where u is a trivial fibration and v is a fibration. By Axiom M5 for C^J (Proposition (3.2.27)) and Propositions (3.2.7) and (3.2.17), v is a trivial cofibration. The required lifting properties follow easily.

Conversely, arguments similar to the proof of Proposition (3.2.18) show that maps u and v with the given lifting properties are weak equivalences. Hence f = uv is a weak equivalence by Axiom M5 for C^J. Details are omitted. □

This completes the proof of Theorem (3.2.2).

§ 3.3. The homotopy theory of pro − C.

Let C be a closed model category which satisfies Condition N (§2.3). We shall show that pro − C inherits a natural closed model structure from the closed model categories C^J (J is a cofinite strongly directed set); this will yield the required homotopy category Ho(pro −C) (see §2.3). One of our main tools is the Mardešić trick (Theorem (2.1.6)) which states that any inverse system is isomorphic to an inverse system indexed by a cofinite strongly directed set.

(3.3.1) <u>Definitions.</u> A map f in pro-C is called a <u>strong cofibration</u> if f is the image in pro-C of a (level) cofibration f_j in some C^J, where J is a cofinite strongly directed set. <u>Strong fibrations</u>, <u>strong trivial cofibrations</u>, and <u>strong trivial fibrations</u> are defined similarly.

A map in pro-C is a <u>cofibration</u> if it is the retract in Maps (pro-C) of a strong cofibration. <u>Fibrations</u>, <u>trivial cofibrations</u>, and <u>trivial fibrations</u> in pro-C are defined similarly. A map f in pro-C is a <u>weak equivalence</u> if f = pi where p is a trivial fibration and i is a trivial cofibration.

(3.3.2) <u>Remarks</u>. Clearly, trivial cofibrations (resp., trivial fibrations) in pro-SS are both cofibrations (resp., fibrations) and equivalences. Corollary (3.3.13), below, shows that the converse assertions hold.

By Definitions (3.3.1) the classes of cofibrations, fibrations, trivial cofibrations, and trivial fibrations are each closed under the formation of retracts. We need to show that the class of weak equivalences is also closed under retracts. Definitions (3.3.1) yield an apparently larger class of weak equivalences than the class of retracts of the (level) weak equivalences in the categories C^J (Definitions (3.2.1)). Definitions (3.3.1) are essentially forced by the requirement that the composition of two weak equivalences yields a weak equivalence (Axiom M5). We do not know if every weak equivalence in pro-C is a retract of a (level) weak equivalence in some C^J.

In this section we shall prove the following.

(3.3.3) <u>Theorem</u>. pro-C, together with the above structure, is a closed model category.

(3.3.4) <u>Theorem</u>. The constant diagram functor $C \longrightarrow \text{pro-}C$ preserves cofibrations, fibrations, and weak equivalences. The inverse limit functor $\lim : \text{pro-}C \longrightarrow C$ preserves fibrations and trivial fibrations.

<u>Proof</u>. Immediate from the Definitions (3.1.1) and Theorem (3.3.3). □

The category of towers, tow - C, inherits a natural closed model structure from pro - C.

The proof of Theorem (3.3.3) involves the following main steps:

Verification of Axiom M0 (Proposition (3.3.5));

Verification of Axiom M2 (Proposition (3.3.8));

Verification of Axiom M6 (Proposition (3.3.9), (3.3.15), and (3.3.17));

Verification of Axiom M5:

Special cases (Propositions (3.3.18), and (3.3.26));

General case (Proposition (3.3.35)).

(3.3.5) <u>Proposition</u> (Verification of Axiom M0). Pro - C admits finite colimits and limits.

<u>Proof</u>. Let Δ be a finite diagram in pro - C. We shall show that Δ has a colimit; the construction of a limit for Δ is similar and omitted. By inserting identity maps if necessary, we may assume that Δ has no loops (the colimits of the original and new diagrams will be isomorphic). Applying the Artin-Mazur reindexing (see Proposition (2.1.5)) to Δ yields an inverse system $\{\Delta_i\}$ of diagrams of C which determines a diagram in pro - C isomorphic to Δ. Applying the Mardešić construction (Theorem (2.1.6)) to $\{\Delta_i\}$ yields a diagram Δ'

over some level category C^J indexed by a cofinite strongly directed set J; further Δ' and Δ are isomorphic diagrams over pro-C. Now let $X' = \text{colim}\ \Delta'$ in C^J (X' is defined levelwise by Proposition (3.2.6)).

We shall now check that $X' = \text{colim}\ \Delta$ in pro-C. Clearly there exists a coherent family of maps from the objects of Δ ($\cong \Delta'$) to X'. To check the universality of X', suppose that there exists a coherent family of maps from Δ to an object Y in pro-C. Let $\tilde{\Delta}$ be the diagram consisting of Δ, Y, and the maps from Δ to Y in pro-C. As above, we may define a diagram $\tilde{\Delta}''$ in some level category C^K indexed by a cofinite strongly directed set K with $\tilde{\Delta}'' \cong \tilde{\Delta}$ in pro-C. Because any diagram which represents $\tilde{\Delta}$ in the Artin-Mazur reindexing process yields a diagram which represents Δ by restriction, there is a natural cofinal functor $T: K \longrightarrow J$. Now let Δ'' and Y'' be the appropriate restrictions of $\tilde{\Delta}''$. Let X'' be the colimit of Δ'' in C^K. Then the maps from Δ'' to Y'' factor uniquely through X''. Further, $\Delta'' \cong T^* \Delta' \cong \Delta' \cong \Delta$ and $X'' \cong T^* X' \cong X'$ in pro-C. Hence X' has the required universal property. □

Artin and Mazur give a non-constructive proof of the following more general result.

(3.3.6) <u>Proposition</u>. [A-M, Propositions A.4.3 and A.4.4]. Let U be a universe such that SS is U-small. Then pro-SS admits U-small colimits and limits. □

(3.3.7) <u>Remarks</u>. Suppose that C admits arbitrary colimits. Let Δ be an <u>infinite</u> diagram over some level category C^J, and X be its colimit in C^J (X is defined levelwise as in Proposition (3.2.6)). It is easy to see that in general X is <u>not</u> the colimit of Δ in pro-C.

The statement of the following proposition is a technical reformulation of Axiom M2; see Remarks (3.3.2).

(3.3.8) <u>Proposition</u>. (Verification of Axiom M2). Any map f in pro-C may be factored as $f = pi$ where i is a trivial cofibration and p is a fibration, or i is a cofibration and p is a trivial fibration.

<u>Proof</u>. By Propositions (2.1.5) and (2.1.6) we may factor f as the composite

$$X \xrightarrow{\cong} X' \xrightarrow{f'} Y' \xrightarrow{\cong} Y,$$

where f' is a level map indexed by a cofinite strongly directed set J. Consider f' as a map in C^J. By Axiom M2 for C^J (Proposition (3.2.24)), we may factor f as $p'i'$ in C^J where p' and i' have the required properties. To complete the proof, let i and p be the respective composite mappings

$$X \xrightarrow{\cong} X' \xrightarrow{i'} Z', \quad Z' \xrightarrow{p'} Y' \xrightarrow{\cong} Y. \quad \square$$

We shall now begin the verification of Axiom M6 for pro-C. The following proposition is a special case of Axiom M1 for pro-C.

(3.3.9) <u>Proposition</u>. Given any commutative solid-arrow diagram

(3.3.10)

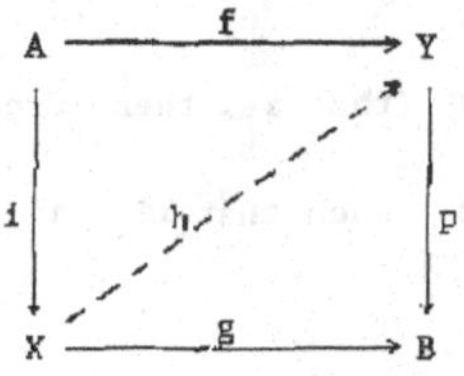

in which either i is a trivial cofibration and p is a fibration, or i is a cofibration and p is a trivial fibration, there exists a filler h.

<u>Proof</u>. We shall only discuss the case in which i is a trivial cofibration. The proof of the other case is similar and omitted.

Since the lifting property described in diagram (3.3.10) is preserved under the formation of retracts, we may assume that i is a strong trivial cofibration indexed by a cofinite strongly directed set J and that p is a strong fibration indexed by a cofinite strongly directed set K.

We shall now replace diagram (3.3.10) by a suitable level diagram. We cannot merely apply the reindexing techniques of §2.1 since it appears unlikely that a cofinal functor $T:L \longrightarrow K$ maps fibrations in C^K into fibrations in C^L (fibrations are <u>not</u> defined levelwise).

Since J is cofinite, we may inductively define an order preserving function $J \longrightarrow K$ $(j \longmapsto k(j))$ and commutative solid-arrow diagrams

(3.3.11)

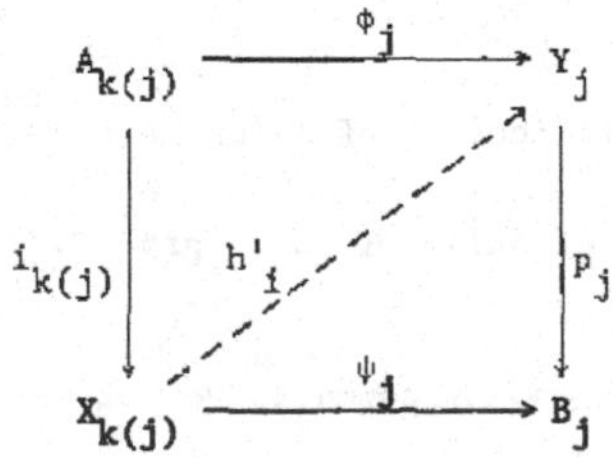

which represent diagram (3.3.10) (that is, there are maps of diagram (3.3.10) to diagrams (3.3.11) over pro - C such that if $j' > j$ there is a commutative diagram of diagrams

$$\begin{array}{ccc} (3.3.10) & \longrightarrow & (3.3.11),\ \text{level } j' \\ & \searrow & \downarrow \\ & & (3.3.11),\ \text{level } j . \end{array}$$

We thus obtain a commutative solid-arrow diagram

$$(3.3.12)\qquad \begin{array}{ccccc} \{A_k\} & \longrightarrow & \{A_{k(j)}\} & \xrightarrow{\{\phi_j\}} & Y_j \\ {\scriptstyle \{i_k\}}\downarrow & & {\scriptstyle \{i_{k(j)}\}}\downarrow & {\scriptstyle \{h'_j\}}\nearrow & \downarrow{\scriptstyle \{p_j\}} \\ \{X_k\} & \longrightarrow & \{X_{k(j)}\} & \longrightarrow & \{B_j\} \end{array}$$

with the following properties: the composites along the top and bottom rows are f and g respectively, and the right-hand square is a level diagram indexed by J (that is, a diagram in C^J). <u>Note</u>: in general the function $J \longrightarrow K$ need not be cofinal, hence $\{A_k\} \neq \{A_{k(j)}\}$, etc. Further, since trivial cofibrations in C^K and C^J are defined levelwise, $\{i_{k(j)}\}$ is a trivial cofibration in C^J.

Since, by hypothesis, $\{p_j\}$ is a fibration in C^J, by Axiom M6 for C^J (Proposition (3.2.28)), there exists a filler $\{h'_j\}$ in diagram (3.3.12). It is clear that the composite mapping

$$\{X_k\} \longrightarrow \{X_{k(j)}\} \xrightarrow{\{h'_j\}} \{Y_j\}$$

is the required filler in diagram (3.3.10). □

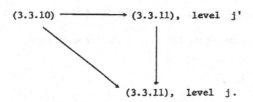

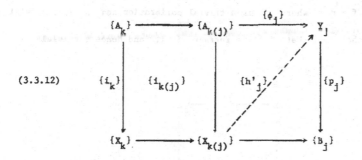

(3.3.13) <u>Corollary</u>. A map is a trivial cofibration if and only if it is both a cofibration and a weak equivalence. A similar description holds for trivial fibrations.

<u>Proof</u>. The "only if" assertions hold by definition. Conversely, let $f:X \longrightarrow Y$ be both a cofibration and a weak equivalence. Using Definitions (3.3.1), write $f = pi$ where i is a trivial cofibration and p is a trivial fibration. We shall see that f is a retract of i, and hence a trivial cofibration.

(3.3.14)

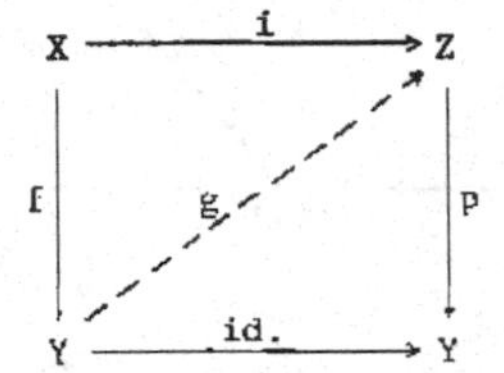

Proposition (3.3.9) yields a filler g in diagram (3.3.14). Rewriting diagram (3.3.14) in the form

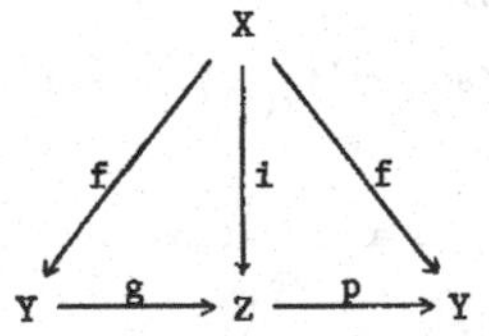

where $pg = id_Y$, shows that f is a retract of i.

Verification of the assertion about fibrations is similar and omitted. □

This answers one of the questions raised in Remarks (3.3.2).

(3.3.15) Proposition. (Verification of Axioms M6a and M6b).

a) A map p is a fibration if and only if for all maps i which are cofibrations and equivalences, the pair (i,p) has the lifting property.

b) A map i is a cofibration if and only if for all maps p which are fibrations and equivalences, the pair (i,p) has the lifting property.

Proof.

a) The "only if" part follows from Corollary (3.3.13) (which shows that i is a trivial cofibration) and Proposition (3.3.9). Conversely, let p be a map with the lifting property of hypothesis a). Use Axiom M2 (Proposition 3.3.8) to write $f = uv$, where u is a fibration and v is a trivial cofibration. As in the proof of Corollary (3.3.13), it follows that p is a retract of u, and hence a fibration.

b) The proof is similar to the proof of a) and is omitted. □

Similar arguments yield the following.

(3.3.16) Proposition. A map i is a trivial cofibration (cofibration and equivalence) if and only if for all fibrations p, the pair (i,p) has the lifting property.

A map p is a trivial fibration (fibration and equivalence) if and only if for all cofibrations i, the pair (i,p) has the lifting property. □

(3.3.17) Proposition. (Verification of Axiom M6c). A map f is a weak equivalence if and only if $f = uv$ where for all cofibrations i and fibrations p, the pairs (i,u) and (v,p) have the lifting property.

Proof. By Proposition (3.3.16), the above characterization of weak equivalences is equivalent to that of Definitions (3.3.1). □

We have completed the verification of Axiom M6 for pro-C, and shall now begin the verification of Axiom M5: weak equivalence is a congruence. This relatively lengthy process consists of first using the lifting properties developed above to verify Axiom M5 under the further assumption that all maps are cofibrations or all maps are fibrations. Secondly, we use the factorizations given by Axiom M2 to verify the general case of Axiom M5 for pro-C.

(3.3.18) Proposition. Suppose that the maps $f:X \longrightarrow Y$ and $g:Y \longrightarrow Z$ are cofibrations. If any two of the maps f, g, and gf are weak equivalences, then so is the third map.

Proof. There are three cases. For Cases I we use Corollary (3.3.13) and Proposition (3.3.9) to characterize maps which are both cofibrations and weak equivalences by their lifting properties.

Case I: Let f and g be weak equivalences. Then for all fibrations p, the pairs (f,p) and (g,p) have the lifting property. Consequently, their composite has the same lifting property; thus gf is a weak equivalence, as required.

Case II: Let f and gf be weak equivalences. We shall show that g is a weak equivalence by verifying that for all fibrations p, the pair (g,p) has the lifting property. Consider a commutative solid-arrow diagram of the form

(3.3.19)

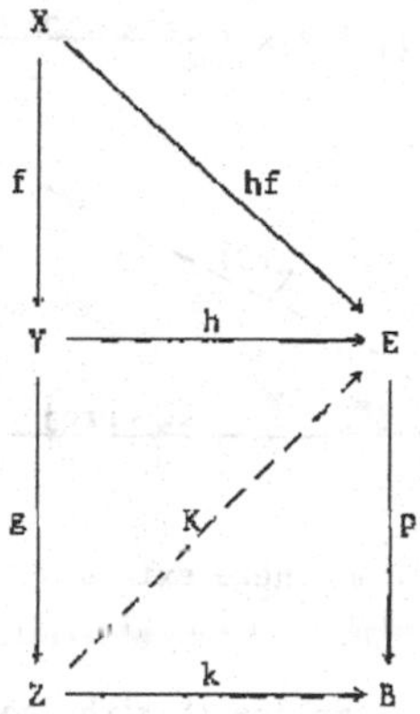

in which p is a fibration. We shall show that g is a trivial cofibration, as required, by constructing a filler K, above.

Because the composite map gf is a trivial cofibration, there is a map $K':Z \longrightarrow E$ such that $K'gf = hf$ and $pK' = k$. <u>Caution</u>: in general $K'g \neq h$. We shall deform K' into the required filler $K:Z \longrightarrow E$.

Because f is a retract of a trivial cofibration $f':X' \longrightarrow Y'$ in some level category C^J, where J is a cofinite strongly directed set, Proposition (2.3.5) implies that f' induces a trivial cofibration

$$i':Y' \times 0 \cup X' \times [0,1] \cup Y' \times 1 \longrightarrow Y' \times [0,1]$$

in C^J. Hence f induces a trivial cofibration

$$i:Y \times 0 \cup X \times [0,1] \cup Y \times 1 \longrightarrow Y \times [0,1]$$

in pro-C. Form the solid-arrow commutative diagram

$$\begin{array}{ccc} Y \times 0 \cup X \times [0,1] \cup Y \times 1 & \xrightarrow{K'g \,\cup\, hg \circ \mathrm{proj} \,\cup\, h} & E \\ {\scriptstyle i}\downarrow & \overset{K^{(2)}}{\nearrow} & \downarrow{\scriptstyle p} \\ Y \times [0,1] & \xrightarrow{kg \circ \mathrm{proj}} & B \end{array}$$

By Axiom M6 (see Proposition (3.3.9)), there exists a filler $K^{(2)}$ above.

As above, Proposition (3.2.29) implies that the cofibration $g:Y \longrightarrow Z$ induces a trivial cofibration

$$i':Z \times 0 \cup Y \times [0,1] \longrightarrow Z \times [0,1].$$

Now form the commutative solid-arrow diagram

$$\begin{array}{ccc} Z \times 0 \cup Y \times [0,1] & \xrightarrow{K' \cup K^{(2)}} & E \\ {\scriptstyle i'}\downarrow & \overset{K^{(3)}}{\nearrow} & \downarrow{\scriptstyle p} \\ Z \times [0,1] & \xrightarrow{K \circ \mathrm{proj}} & B \end{array}$$

Again, as above, there exists a filler $K^{(3)}$.

Finally, let K be the composite mapping

$$Z \cong Z \times 1 \longrightarrow Z \times I \xrightarrow{K^{(3)}} E.$$

Then K is the required filler in Diagram (3.3.19) (easy check), so g is an equivalence, as required.

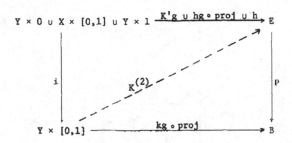

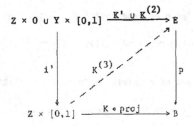

<u>Case III</u>: Let g and gf be weak equivalences. Assume, for now, the following lemma.

(3.3.20) <u>Lemma</u>. Suppose that a map i in pro - C has the left-lifting-property with respect to all fibrations of fibrant objects. Then i is a trivial cofibration.

Consider a solid-arrow commutative diagram

(3.3.21)

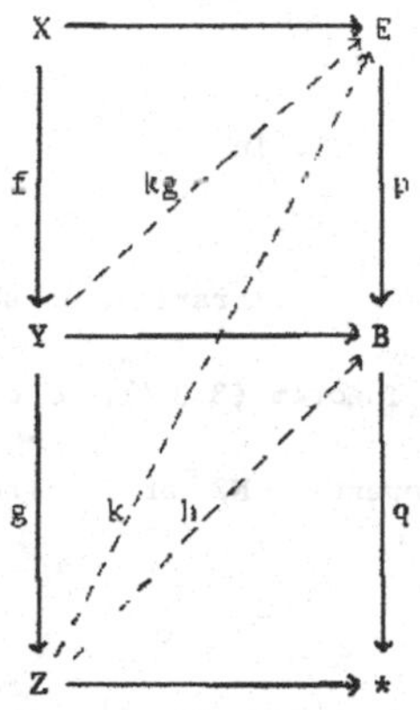

in which p is a fibration, and B is fibrant (that is, q is a fibration). Because g and gf are weak equivalences we may successively construct the fillers h and kg above. Because kg is a filler in the top square of Diagram (3.3.21) above, f has the required lifting property to be a trivial cofibration (Lemma (3.3.20)). Hence, f is a weak equivalence, as required. □

(3.3.22) <u>Proof of Lemma 3.3.20</u>. Let L denote the class of all p for which the pair (i,p) has the lifting property. It is easy to check that L has the following three properties.

a) A pullback of a map in L is in L .

b) Let $E(j)$ be an inverse system of objects in pro - C indexed by a cofinite directed set J with a least element 0. If all the induced maps

$$E(j) \longrightarrow \lim_{k<j}\{E(k)\}, \quad j \in J$$

are in $\mathcal{L}$, then the induced map

$$\lim_J\{E(j)\} \longrightarrow E(0)$$

is in $\mathcal{L}$.

c) A retract of a map in $\mathcal{L}$ is in $\mathcal{L}$.

To show that i is a trivial cofibration, we shall show that $\mathcal{L}$ contains all fibrations in pro - C. By Theorem (3.3.4), $\mathcal{L}$ contains all fibrations of fibrant objects in C; by property N2 of C (see §2.3), $\mathcal{L}$ contains all fibrations in C.

Now let $p: E \longrightarrow B$ be a fibration in C^J, where J is a cofinite strongly directed set. Let $J^* = J \cup \{0\}$, where $0 < j$ for all j in J. Define an inverse system $\{E(j) \mid j \in J^*\}$ over pro - C as follows. Set $E(0) = B$. For $j \in J$, define $E(j)$ by the pullback diagram

(3.2.23)
$$\begin{array}{ccc} E(j) & \longrightarrow & E_j \\ \downarrow & & \downarrow \\ B & \longrightarrow & B_j \end{array}$$

in pro - C. That is, for $k \leqq j$, $E(j)_k = E_k$, otherwise, $E(j)_k$ is a suitable pullback. Because J is a cofinite strongly directed set, $\lim \{E(j)\} \cong E$. We shall see that $\{E(j)\}$ satisfies the hypothesis of property (b) above.

For now, consider a fixed j in J. To show that the map

$$E(j) \longrightarrow \lim_{k<j}\{E(k)\}$$

is in L, we shall show that given a commutative solid arrow diagram of the form

$$\begin{array}{ccc} A & \xrightarrow{h_1} & E(j) \\ {\scriptstyle i}\downarrow & {\scriptstyle H} \nearrow & \downarrow \\ X & \xrightarrow[h_2]{} & \lim_{k<j} E(k) \end{array}$$

$$H : X \dashrightarrow E(j)$$

in pro-C, there exists a filler H. To do this, form the pullback diagram in pro-C

$$\begin{array}{ccc} \lim_{k<j} E(k) & \longrightarrow & \lim_{k<j} E_k \\ \downarrow & & \downarrow \\ B & \longrightarrow & \lim_{k<j}\{B_k\}. \end{array}$$

Thus h_2 corresponds to maps

$$h_2' : X \longrightarrow \lim_{k<j}\{E_k\}, \quad \text{and}$$

$$h_2'' : X \longrightarrow B$$

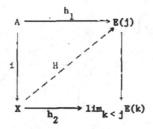

whose images in $\lim_{k<j}\{B_k\}$ agree. Also, because $E(j)$ is defined by the pullback diagram (3.2.23) in pro-C, h_1 corresponds to the pair of maps

$$h_1':A \longrightarrow E_j, \quad \text{and}$$

$$h_1'':A \longrightarrow B,$$

whose images in B_j are equal. Now consider the resulting commutative solid-arrow diagram

(3.3.25)

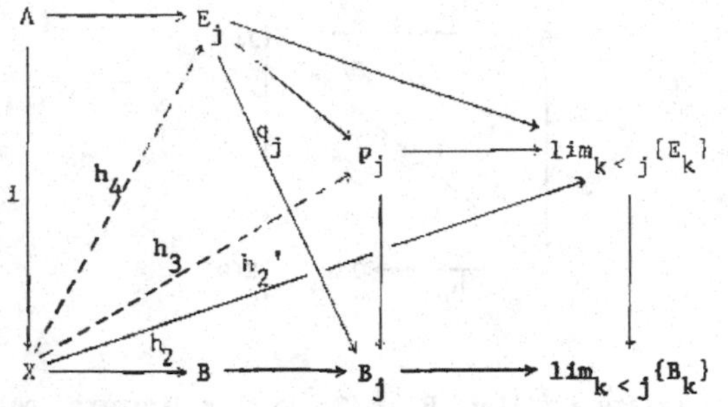

Here P_j is a pullback, and the composite mapping $A \longrightarrow X \longrightarrow B$ is h_1''. By the universal property of pullbacks, there is a unique filler h_3. By Proposition (3.2.7), q_j is a fibration in C, hence $q_j \in L$, so that there exists a filler h_4 above. Diagrams (3.2.23) and (3.2.25) show that the maps $h_2'':X \longrightarrow B$ and $h_4:X \longrightarrow E_j$ induce a unique map $H:X \longrightarrow E(j)$, which is the required filler in diagram (3.2.24). We have shown that the inverse system $E(j)$ satisfies the hypothesis of condition (b) above. Hence the induced map

$$p:E = \lim\{E(j)\} \longrightarrow E(0) = B$$

is in L. Therefore L contains all strong fibrations in pro-C. By property (c) above, L contains all fibrations in pro-C. Therefore i is a trivial cofibration, as required. □

Similar techniques yield the following.

(3.3.26) Proposition. Let $f:X \to Y$, and $g:Y \to Z$ be fibrations. If any two of the maps f, g, and gf are weak equivalences, so is the third.

The proof is analogous to the proof of Proposition (3.3.18) and requires a lemma "dual" to Lemma (3.3.20).

(3.3.27) Lemma. Suppose that a map p in pro-C has the right-lifting-property with respect to all cofibrations of cofibrant objects. Then p is a trivial fibration.

The proof is similar to the proof (3.3.22), and somewhat simpler because cofibrations in C^J are defined levelwise. Details of the proofs of the above Proposition and Lemma are omitted.

We now begin the proof of the general case of Axiom M5 with four preliminary lemmas.

(3.3.28) Lemma. Let $f:X \to Y$ be a weak equivalence. Suppose that $f = pi$, where i is a trivial cofibration and p is a fibration. Then p is a trivial fibration.

Proof. By Definitions (3.3.1), we may write $f = p'i'$, where i' is a trivial cofibration and p' is a trivial fibration. Form the commutative solid-arrow diagram

(3.3.29)

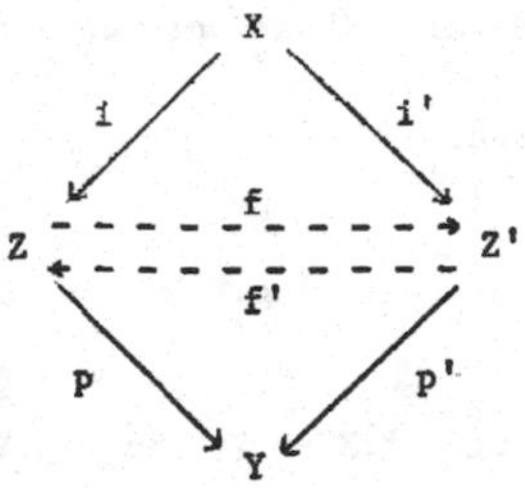

By Axiom M6a (Proposition (3.3.15)), there exist maps f and f' (as shown above) such that diagram (3.3.29) together with <u>either</u> dotted arrow commutes. Hence $f'fi = i$ and $pf'f = 0$. Thus there exists a commutative solid-arrow diagram (see Proposition (2.3.5))

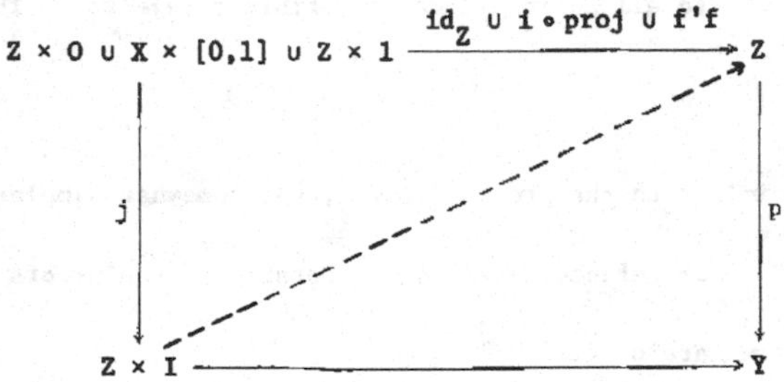

As in the proof of Case II of Proposition (3.3.18), j is a trivial cofibration. Hence there exists a filler H, that is, a homotopy $H: id_Z \simeq f'f$ relative to X which covers id_Y. A similar construction yields a homotopy $H': id_{Z'} \simeq ff'$. Therefore the fibrations p and p' are fiber-homotopy-equivalent. Hence they have similar lifting properties (use deformations analogous to those in the proof of Case II of Proposition (3.3.18)), so that p is a trivial fibration by Proposition (3.3.16). $\square$

(3.3.30) Lemma. Let $f:X \longrightarrow Y$ be a weak equivalence. Suppose that $f = pi$, where p is a trivial fibration and i is a cofibration. Then i is a trivial cofibration.

The proof is similar to that of Lemma (3.3.28), and is omitted.

At this point we recall Condition N3 for C: at least one of the following statements holds.

N3a. All objects of C are cofibrant.

N3b. All objects of C are fibrant.

We shall assume N3a for the remainder of this section, unless otherwise specified. If instead N3b holds, replace "fibration" by "cofibration" in Lemma (3.3.31), below, "dualize" the proof of Lemma (3.3.32), below, and make some other obvious changes. Details are omitted.

(3.3.31) Lemma. Let $p:E \longrightarrow B$ be a trivial fibration. Then there exists a section $s:B \longrightarrow E$ (that is, $ps = id_B$); further, any section is a trivial cofibration.

Proof. By Assumption N3, B is cofibrant, so Axiom M6 (see Proposition (3.3.9)) yields a filler s in the commutative solid-arrow diagram

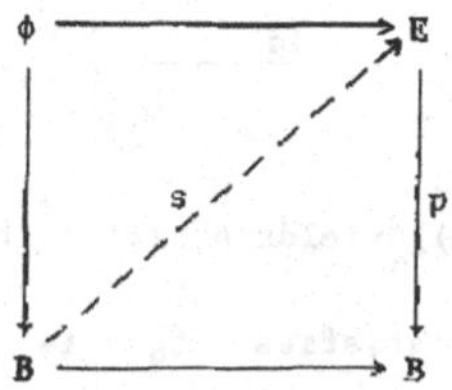

Now, let $s':B \to E$ be a section to p. As in the proof of Lemma (3.3.28), there exists a homotopy $H:E \times [0,1] \to E$ over id_B from sp to id_E. As in the proof of Proposition (3.3.18), Case II, or the proof of [Q -1, Lemma I.5.1, §4], we may use H to show that s' is a trivial cofibration. □

(3.3.32) Lemma. Let $f:X \to Y$ be a trivial fibration and $g:Y \to Z$ be a trivial cofibration. Then the composite gf is a weak equivalence.

Proof. By Axiom M2 (Proposition (3.3.8)) we may factor gf as pi, where i is a cofibration and p is a trivial fibration. Form the commutative solid-arrow diagram

(3.3.33)

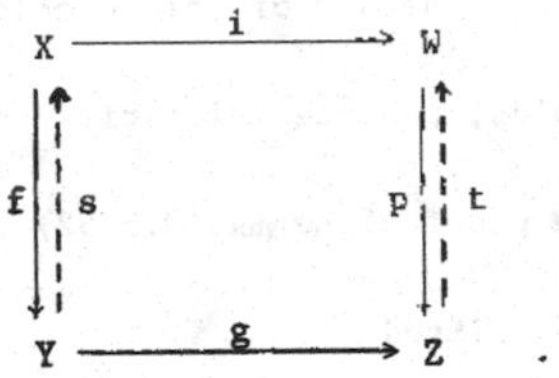

Lemma (3.3.31) yields a section s to f. Because $pis = gfs = g$, there exists a commutative solid-arrow diagram

(3.3.34)

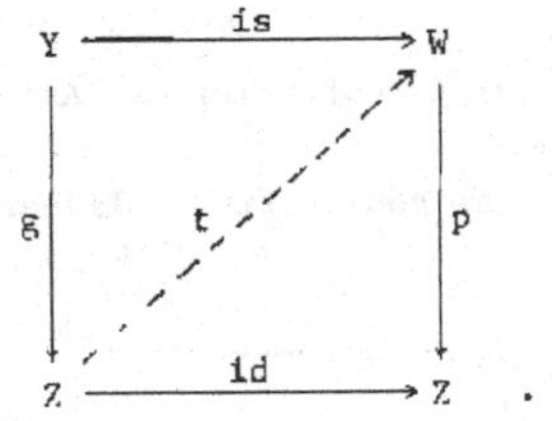

Axiom M6 (see Proposition (3.3.9)) yields a filler t in diagram (3.3.34). The map t is a section to p and satisfies $tg = is$ (see diagram (3.3.33)).

Axiom M5 <u>for cofibrations</u> (Proposition (3.3.18)) implies that $tg(= is)$ and i are trivial cofibrations. Hence gf $(= pi)$ is a weak equivalence (Definitions (3.3.1)), as required. □

(3.3.35) <u>Proposition</u>. (Verification of Axiom M5). Let

$$X \xrightarrow{f} Y \xrightarrow{g} Z$$

be a diagram in which two of the maps f, g, and gf are weak equivalences. Then so is the third map.

<u>Proof</u>. There are three cases.

<u>Case I</u>: Let f and g be weak equivalences. As in Definitions (3.3.1), write $f = pi$ and $g = pj$, where p and q are trivial fibrations and i and j are trivial cofibrations. By Lemma (3.3.32) the composite mapping jp is a weak equivalence, write $jp = rk$, where r is a trivial fibration and k is a trivial cofibration. By Propositions (3.3.26) and (3.3.18) respectively, the composite mapping qr is a trivial fibration and the composite mapping ki is a trivial cofibration. Hence gf $(= qrki)$ is a weak equivalence, as required.

<u>Case II</u>. Let f and gf be weak equivalences. Write $f = pi$ and $g = qj$, where p is a trivial fibration, i and j are trivial cofibrations, and q is a fibration. Then $gf = qjpi$. By Lemma (3.3.32), jp is a weak equivalence. We may therefore write $jp = rk$, where r is a trivial fibration and k is a trivial cofibration. We have thus factored the weak equivalence gf as $(qr)(ki)$, where ki is a trivial fibration (use Proposition (3.3.18)) and qr is a fibration (by Axiom M6a (Proposition (3.3.15)) which implies that the class

of fibrations is closed under composition). By Lemma (3.3.28), qr is a trivial fibration. Finally, Proposition (3.3.26) implies that q is a trivial fibration, so that $g(=qj)$ is a weak equivalence as required.

<u>Case III</u>. Let g and gf be weak equivalences. As in Case II, factor g and f so that $gf = qjpi$, where j is a trivial cofibration, p and q are trivial fibrations, and i is a cofibration. Proceed as in Case II, using Lemmas (3.3.32) and (3.3.30). Details are omitted. □

This completes the proof that pro - C is a closed model category.

We shall conclude this section by describing cofibrations and trivial cofibrations up to isomorphism. This extends J. Grossman's [Gros -1, §4] characterization of cofibrations in his closed model structure on tow - SS.

(3.3.36) <u>Proposition</u>. A map f in pro - C is a cofibration (respectively, trivial cofibration) if and only if f is isomorphic to a strong cofibration (respectively, strong trivial cofibration) (in some C^J).

<u>Proof</u>. The "only if" part is obvious. For the "if" part, first reindex f if necessary (Proposition (3.1.4) and Theorem (2.1.6)) so that $f = \{f_j\}:\{X_j\} \longrightarrow \{Y_j\}$, $j \in J$, with J cofinite. By Definitions (3.3.1), $\{f_j\}$ is a retract of a strong cofibration (respectively, strong trivial cofibration) $\{f'_k\}:\{X'_k\} \longrightarrow \{Y'_k\}$, $k \in K$, with K cofinite. Form the following commutative diagram in pro - C

(3.3.37)

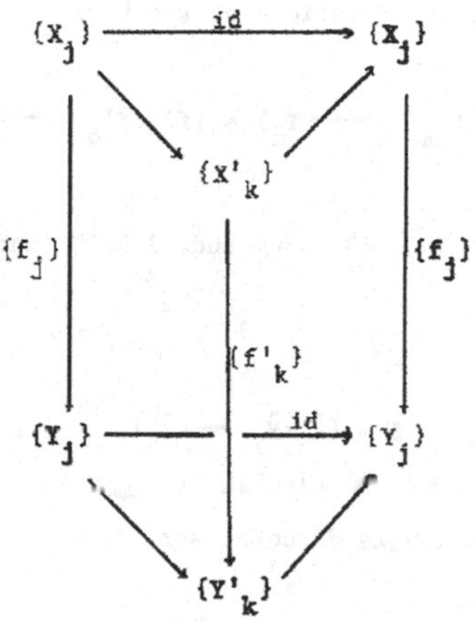

We shall use diagram (3.3.37) to construct a level cofibration (respectively, level trivial cofibration) f'' isomorphic to f. First consider the left front square of diagram (3.3.37). We shall say that

(3.3.38) $$(f'_\ell : X'_\ell \longrightarrow Y'_\ell) < (f_m : X_m \longrightarrow Y_m)$$

if the square (shown in perspective)

(3.3.39)

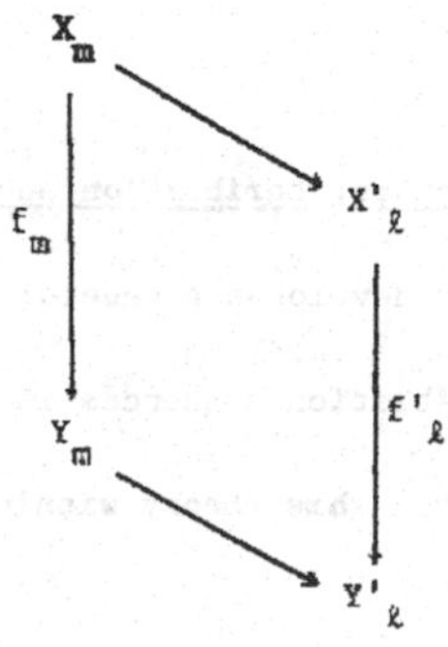

is a left front square of diagram (3.3.37). Similarly, use the right front square of diagram (3.3.37) to define relations of the form

$$(3.3.40) \qquad (f_m : X_m \longrightarrow Y_m) < (f'_n : X'_n \longrightarrow Y'_n).$$

By diagram (3.3.37), relations (3.3.38) and (3.3.40) and their composites yield an inverse system of maps

$$\tilde{f} = \{\tilde{f}_k : \tilde{X}_k \longrightarrow \tilde{Y}_k\}, \quad k \in K$$

indexed by a (cofinite) strongly directed set K.

Further, $\{f'_j : X'_j \longrightarrow Y'_j\}$, $j \in J$, <u>together with bonding maps</u> induced by diagrams (3.3.37) - (3.3.40) is a cofibration (resp., trivial cofibration) and is cofinal in $\tilde{f}$, hence isomorphic to f. Also, $\{f_i : X_i \longrightarrow Y_i\}$ admits a cofinal subsystem which is cofinal in $\tilde{f}$ (the bonding maps agree by diagrams (3.3.37) - (3.3.40)). The conclusion follows. □

(3.3.41) <u>Remarks</u>. The above proof used the fact that cofibrations in C^J are defined levelwise. We do not know whether the analogue of Proposition (3.3.36) for fibrations holds.

§3.4. <u>Suspension and loop functors, cofibration and fibration sequences</u>.

D. Quillen [Q-1, §§I.2-3] developed a general theory of suspension and loop functors and cofibration and fibration sequences in Ho(C), where C is a closed model category. We shall sketch this theory within the context of our closed model structures on pro-C.

(3.4.1) <u>Morphisms in</u> <u>Ho(pro - C)</u>. Let X be a cofibrant object and Y be a fibrant object in pro - C. Then

$$\mathrm{Ho}(\text{pro-}C)(X,Y) = [X,Y],$$

where $[X,Y]$ denotes the set of homotopy classes of maps from X to Y (in pro - C) with respect to the cylinder $X \otimes [0,1]$. There is another dual description of Ho(pro - C)(X,Y). Factoring the diagonal map $Y \longrightarrow Y \times Y$ as the composite of a trivial cofibration followed by a fibration

$$Y \longrightarrow Y^{[0,1]} \longrightarrow Y \times Y$$

yields the cocylinder $Y^{[0,1]}$. <u>Caution</u>: in general $Y^{[0,1]}$ need <u>not</u> depend functorially upon Y (compare with Definition (2.3.4)). One can easily show that maps $f,g: X \rightrightarrows Y$ are homotopic if and only if there exists a commutative diagram

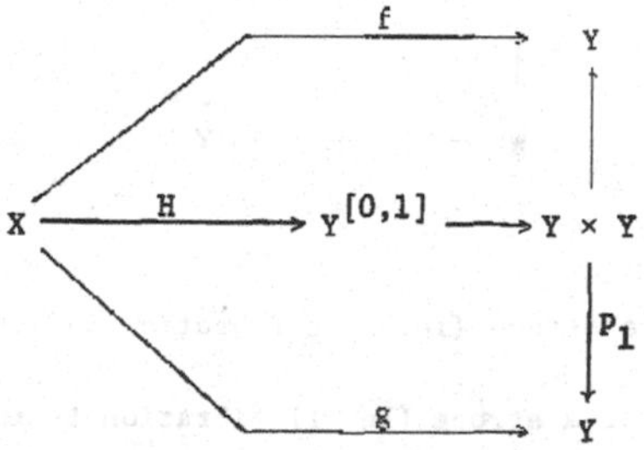

here p_0 and p_1 denote the projections onto the first (Y^0) and second (Y^1) factors in $Y \times Y$, respectively.

Now, let C_* be a <u>pointed</u> closed model category, that is, a closed model category which is also a pointed category. Then pro - C_* becomes a pointed closed model category (the point $*$ of C_* is also the point of pro - C_*). We shall follow the "usual" conventions and write $\vee$ for the sum (coproduct) in pro - C_*.

(3.4.2) Definitions. If $f:X \longrightarrow Y$ is a cofibration in pro - C_* we shall write $* \vee_X Y$ for the cofibre of f defined by the pushout diagram

$$\begin{array}{ccc} X & \longrightarrow & * \\ {\scriptstyle f}\downarrow & & \downarrow \\ Y & \longrightarrow & * \vee_X Y \, . \end{array}$$

If $f:X \longrightarrow Y$ is a fibration in pro - C_*, we shall write $* \times_Y X$ for the fibre of f defined by the pullback diagram

$$\begin{array}{ccc} * \times_Y X & \longrightarrow & X \\ \downarrow & & \downarrow \\ * & \longrightarrow & Y \end{array}$$

Note that the cofibre of a strong (level) cofibration is just the levelwise cofibre; similarly the fibre of a strong (level) fibration is the levelwise fibre.

(3.4.3) Suspensions and loop spaces. Let $X \in$ pro - C_* be cofibrant. Choose a cylinder object $X \times [0,1]$, and let ΣX be the cofibre of the map $i_0 + i_1 : X \vee X \longrightarrow X \vee [0,1]$. We shall call ΣX a suspension of X. Loop spaces are defined dually. Let $Y \in$ pro - C_* be fibrant. Choose a cocylinder object $Y^{[0,1]}$, and let ΩY be the fibre of the map $(p_0,p_1):Y^{[0,1]} \longrightarrow Y \times Y$.

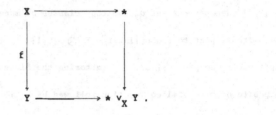

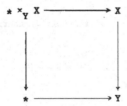

We shall call ΩY a loop-space of Y. Note that ΣX above is cofibrant and ΩY is fibrant. Caution: In general Σ and Ω need not be functors on pro $-C_*$ (see Definition (2.3.4) and paragraph (3.4.1)).

On the other hand, Quillen proves the following theorem for arbitrary closed pointed model categories.

(3.4.4) Theorem [Q - 1, §§I.2]. We may extend Σ and Ω to an adjoint pair of functors on all of $\mathrm{Ho}(\text{pro} - C_*)$:

$$\mathrm{Ho}(\text{pro} -C_*)(\Sigma X,Y) = \mathrm{Ho}(\text{pro} - C_*)(X,\Omega Y).$$

If X is cofibrant and Y is fibrant,

$$[\Sigma X,Y] \cong [X,\Omega Y].$$

Σ and Ω are defined up to canonical isomorphism. Also, $\mathrm{Ho}(\text{pro} - C_*)(\Sigma -,-) = \mathrm{Ho}(\text{pro} - C_*)(-,\Omega -)$ as functors from $(\text{pro} -C_*)^{op} \times (\text{pro} -C_*)$ to groups.

The proof is similar to the proof for the category of pointed spaces, except that somewhat more care is needed because of the choices involved in defining Σ and Ω. The group structure on $\mathrm{Ho}(\text{pro} -C_*)(\Sigma X,Y)$ comes from a co-H structure $\Sigma X \longrightarrow \Sigma X \vee \Sigma X$ which makes ΣX a cogroup object in $\mathrm{Ho}(\text{pro} -C_*)$; the corresponding group structure on $\mathrm{Ho}(\text{pro} -C_*)(X,\Omega Y)$ comes from an H-structure $\Omega Y \times \Omega Y \longrightarrow \Omega Y$ which makes ΩY a group object in $\mathrm{Ho}(\text{pro} -C_*)$.

(3.4.5) Short cofibration sequences. Let $f:A \longrightarrow X$ be a cofibration in Top_*. Let

$$M_f = A \times [0,1] \cup_A X/* \times [0,1]$$

be the mapping cylinder of f, and consider the induced cofibration

$$A \xrightarrow{i_0} M_f$$

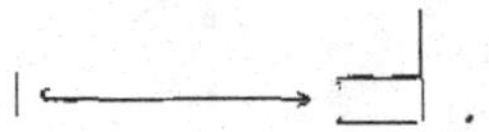

If C_{i_0} denotes the cofibre of i_0, CA the reduced cone on A, and ΣA the reduced suspension of A, we may form the sequence

$$A \xrightarrow{i_0} M_f \longrightarrow C_{i_0}$$

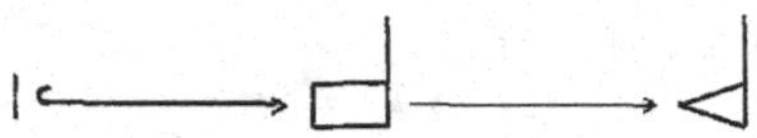

and induced diagram

$$C_{i_0} \xleftarrow{\cong} CA \cup_A M_f \longrightarrow \Sigma A \vee C_{i_0}$$

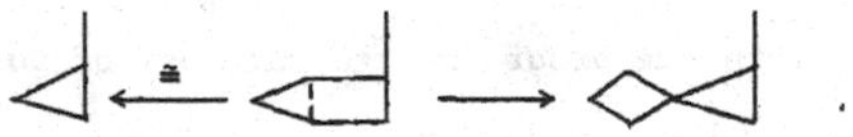

Further, in $\mathrm{Ho}(\mathrm{Top}_*)$ the composite mapping $n: C_{i_0} \longrightarrow \Sigma A \vee C_{i_0}$ is a <u>coaction</u> of the <u>cogroup object</u> (co -H space with a co-inverse) ΣA on C_{i_0}.

We shall extend the above observations to arbitrary cofibrations in $(\text{pro}-C_*)_c$, the full subcategory of cofibrant objects in $\text{pro}-C_*$. (The "cofibrant" condition means that the map $* \longrightarrow X$ is a cofibration.) The following notation is needed. If $A \longrightarrow X$ is a map in $\text{pro}-C_*$ we shall write $A \otimes [0,1] \vee_A X$

for the <u>cofibre sum</u> (pushout)

$$\begin{array}{ccc} A & \longrightarrow & X \\ {\scriptstyle i_1}\downarrow & & \downarrow \\ A \otimes [0,1] & \longrightarrow & A \otimes [0,1] \vee_A X . \end{array}$$

More generally, we shall write $A \otimes [0,1] \vee_A -$ to denote the cofibre sum with $i_1 : A \longrightarrow A \otimes [0,1]$, and $- \vee_A A \otimes [0,1]$ to denote the cofibre sum with $i_0 : A \longrightarrow A \otimes [0,1]$.

Now let $A \rightarrow X$ be a cofibration in $(\text{pro}-C_*)_c$, that is, A is cofibrant in $\text{pro}-C_*$. Let Z be the cofibre of A. We shall define a <u>coaction</u> of the cogroup A on Z. To do this, form the commutative diagram

(3.4.6)
$$\begin{array}{ccc} A \otimes [0,1] \vee_A X & \longrightarrow & * \vee_A A \otimes [0,1] \vee_A X \\ \downarrow & & \downarrow \\ X \otimes [0,1] & \longrightarrow & Z \vee_X X \otimes [0,1] \\ {\scriptstyle i_0}\uparrow & & \uparrow {\scriptstyle i_0'} \\ X & \longrightarrow & Z \end{array}$$

in which the cylinder object $A \otimes [0,1]$ is obtained by factoring the natural map $X \vee_A A \otimes [0,1] \vee_A X \longrightarrow X$ as a cofibration followed by a trivial fibration

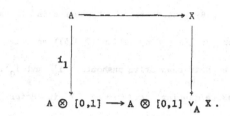

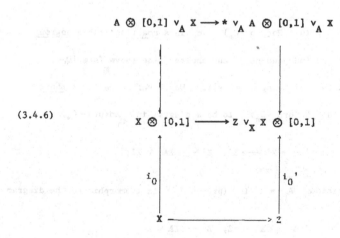

$$X \vee_A A \otimes [0,1] \vee_A X \longrightarrow X \otimes [0,1] \longrightarrow X$$

(see Proposition (2.3.5)). By construction both squares in diagram (3.4.6) are pushout squares. Further, i (see Proposition (2.3.5)) and i_0 are trivial cofibrations, hence so are their respective pushouts i' and i_0'. Diagram (3.4.6) induces the composite mapping $c: Z \longrightarrow \Sigma A \vee Z$ in $\mathrm{Ho}(\text{pro-}C_*)$ defined below.

$$Z \xrightarrow{[i_0]} Z \vee_X X \otimes [0,1]$$

$$\longrightarrow * \vee_A A \otimes [0,1] \vee_A X$$

$$\longrightarrow * \vee_A A \otimes [0,1] \vee_A * \vee Z$$

$$\approx \Sigma A \vee Z$$

It is easy to check that, in $\mathrm{Ho}(\text{pro-}C_*)$, c is a <u>coaction</u> of the <u>cogroup</u> A on Z and that c is independent of any choices made above (see [Q-1, Proposition I.3.1 and the following remarks]). We therefore define a <u>short cofibration sequence</u> in $\mathrm{Ho}(\text{pro-}C_*)$ to be a diagram in $\mathrm{Ho}(\text{pro-}C_*)$

(3.4.7) $$A' \longrightarrow X' \longrightarrow Z', \quad Z' \longrightarrow \Sigma A' \vee Z',$$

which for some cofibration $A \rightarrow X$ in $(\text{pro-}C_*)_c$ is isomorphic to the diagram

(3.4.8) $$A \longrightarrow X \longrightarrow Z, \quad Z \longrightarrow \Sigma A \vee Z$$

constructed above.

(3.4.9) <u>Proposition</u>. A short cofibration sequence

$$A \longrightarrow X \longrightarrow Z, \quad Z \longrightarrow \Sigma A \vee Z$$

induces a short cofibration sequence

$$X \longrightarrow Z \longrightarrow \Sigma A, \quad \Sigma A \longrightarrow \Sigma X \vee \Sigma A,$$

the "connecting map" $Z \longrightarrow \Sigma A$ is the composite

$$Z \longrightarrow \Sigma A \vee Z \longrightarrow \Sigma A,$$

and the coaction $\Sigma A \longrightarrow \Sigma X \vee \Sigma A$ is the composite

$$\Sigma A \longrightarrow \Sigma A \vee \Sigma A \xrightarrow{-id \vee id} \Sigma A \vee \Sigma A \longrightarrow \Sigma X \vee \Sigma A,$$

where $-id:\Sigma A \longrightarrow \Sigma A$ is the inverse in the cogroup ΣA.

The proof is similar to the usual proof of the corresponding assertion in $Ho(Top_*)$, and dual to the proof of [Q -1, Proposition I.3.3]. Details are omitted.

(3.4.10) <u>Long cofibration sequences</u>. A short cofibration sequence $A \to X \to Z$, $Z \longrightarrow \Sigma A \vee Z$ in $Ho(pro-C_*)$ induces a <u>long cofibration sequence</u> (Barratt-Puppe sequence)

$$A^f \to X \to Z \to \Sigma A \to \cdots \to \Sigma^n A \to \Sigma^n X \to \Sigma^n Z \to \Sigma^{n+1} A \to \cdots .$$

Also, for any object Y in $pro-C_*$, the sequence

$$(3.4.11) \qquad \cdots \longrightarrow Ho(pro-C_*)(\Sigma^n Z,Y) \longrightarrow Ho(pro-C_*)(\Sigma^n X,Y)$$

$$\longrightarrow Ho(pro-C_*)(\Sigma^n A,Y) \longrightarrow \cdots$$

$$\longrightarrow Ho(pro-C_*)(\Sigma A,Y) \longrightarrow Ho(pro-C_*)(Z,Y)$$

$$\longrightarrow Ho(pro-C_*)(X,Y) \longrightarrow Ho(pro-C_*)(A,Y)$$

has the usual exactness properties:

a) Sequence (3.4.11) is exact as a sequence of pointed sets, and maps of pointed sets;

b) Sequence (3.4.11) is exact to the left of $Ho(pro-C_*)(\Sigma X,Y)$ as a sequence of groups and homomorphisms;

c) Two maps in $Ho(pro-C_*)(X,Y)$ have the same image in $Ho(pro-C_*)(X,Y)$ if and only if they differ by the action of the group $Ho(pro-C_*)(\Sigma A,Y)$ on $Ho(pro-C_*)(Z,Y)$.

d) Two maps $g_1, g_2 \in Ho(pro-C_*)(A,Y)$ have the same image in $Ho(pro-C_*)(Z,Y)$ if and only if $g_2 = (\Sigma f^*)h \circ g_1$ for some map h in $Ho(pro-C_*)(\Sigma X,Y)$.

As above, compare the usual exactness properties of Barratt-Puppe sequences and [Q-1, Proposition I.3.4'] for the proof.

We now summarize the properties of cofibration sequences.

(3.4.12) <u>Proposition</u>. (dual to [Q-1, Proposition I.3.5]). The class of short cofibration sequences in $Ho(pro-C_*)$ has the following properties:

a) Any map $f: X \longrightarrow Y$ in $\mathrm{Ho}(\text{pro}-C_*)$ may be embedded in a cofibration sequence

$$X \xrightarrow{f} Y \longrightarrow Z, \quad Z \longrightarrow \Sigma X \vee Z.$$

b) Given a commutative solid-arrow diagram

$$(3.4.13) \qquad \begin{array}{ccccc} X & \longrightarrow & Y & \longrightarrow & Z \\ \Big\downarrow f & & \Big\downarrow g & & \Big\downarrow h \\ X' & \longrightarrow & Y' & \longrightarrow & Z' \end{array} \qquad \begin{array}{ccc} Z & \longrightarrow & \Sigma X \vee Z \\ \Big\downarrow h & & \Big\downarrow f \vee h \\ Z' & \longrightarrow & \Sigma X' \vee Z' \end{array}$$

in which the rows are short cofibration sequences, the filler h exists.

c) If the maps f and g in diagram (3.4.13) are weak equivalences, so is the filler h.

d) Proposition (3.4.9) holds.

We omit the proof.

The following straight-forward proposition yields many cofibration sequences in $\mathrm{Ho}(\text{pro}-C_*)$.

(3.4.13) Proposition. Let $\{A_j \longrightarrow X_j \longrightarrow Z_j\}$ be an inverse system of cofibrations of cofibrant objects $A_j \longrightarrow X_j$, with cofibres Z_j, over C_* indexed by a cofinite strongly directed set J. Then there is an induced cofibration sequence

$$\{A_j\} \longrightarrow \{X_j\} \longrightarrow \{Z_j\}, \quad \{Z_j\} \longrightarrow \Sigma\{X_j\} \vee \{Z_j\}$$

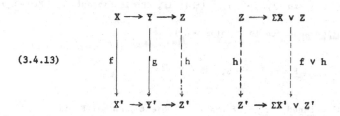

in $\mathrm{Ho(pro-C_*)}$. □

All of the above theory may be dualized to obtain short and long fibration sequences. Quillen [Q-1, §I.3] discusses fibrations explicitly. We shall summarize this discussion below.

A fibration $p:Y \to B$ in $(\mathrm{pro-C_*})_f$ (that is, B is fibrant in $\mathrm{Pro-C_*}$) induces a short fibration sequence in $\mathrm{Ho(pro-C_*)}$,

(3.4.14) $$F \longrightarrow Y \xrightarrow{p} B, \quad \Omega B \times F \xrightarrow{m} F,$$

in which m is a well-defined action of the group object ΩB on F in $\mathrm{Ho(pro-C_*)}$. There is also an induced short fibration sequence

(3.4.15) $$\Omega B \xrightarrow{\partial} F \longrightarrow E, \quad \Omega E \times \Omega B \xrightarrow{n} \Omega B,$$

where ∂ is the composite map

$$\Omega B \xrightarrow{id \times *} \Omega B \times F \xrightarrow{\quad m \quad} F,$$

and n is the composite map

$$\Omega E \times \Omega B \longrightarrow \Omega B \times \Omega B \xrightarrow{-id \times id} \Omega B \times \Omega B \longrightarrow \Omega B.$$

This is [Q-1, Proposition I.3.3], compare Proposition (3.4.9). Hence, there is an induced long fibration sequence

(3.4.16) $$\cdots \longrightarrow \Omega^{n+1}B \xrightarrow{\Omega^n\partial} \Omega^n F \longrightarrow \Omega^n E \longrightarrow \Omega^n B \longrightarrow \cdots$$

$$\longrightarrow B \xrightarrow{\partial} F \longrightarrow E \longrightarrow B$$

with exactness properties analogous to properties (a) - (d) of long cofibration sequences (see Paragraph (3.4.10) above)[Q -1, Proposition I.3.4].

We shall need the following dual of Proposition (3.4.13).

(3.4.17) Proposition. Let $\{F_j \longrightarrow E_j \longrightarrow B_j\}$ be an inverse system of fibrations of fibrant objects E_j and B_j, with fibres F_j over $* \in B_j$, indexed by a cofinite strongly directed set J. Then there is an induced fibration sequence

$$\{F_j\} \longrightarrow \{E_j\} \longrightarrow \{B_j\}, \quad \Omega\{B_j\} \times \{F_j\} \longrightarrow \{F_j\}$$

in $\mathrm{Ho}(\text{pro}-C_*)$.

Proof. We first replace $\{B_j\}$ by a fibrant object $\{B'_j\}$, and next replace the map $\{E_j\} \longrightarrow \{B_j\}$ by a fibration $\{E'_j\} \longrightarrow \{B'_j\}$ in C_*^J. (Recall that the natural functor $C_*^J \longrightarrow \text{pro}-C_*$ preserves the closed model structures.) Caution: recall that fibrations in C_*^J are not defined levelwise, but the fibre of a strong fibration in $\text{pro}-C_*$, that is, a fibration in C_*^J, is just the levelwise fibre. Use the proof of Proposition (3.2.24) (Axiom M2 for C_*^J) first to factor the map $\{B_j\} \longrightarrow *$ as a level trivial cofibration followed by a fibration

$$\{B_j\} \xrightarrow{\{i_j\}} \{B'_j\} \longrightarrow *$$

(this makes B'_j fibrant) and then to factor the composite map

$$\{E_j\} \longrightarrow \{B_j\} \longrightarrow \{B'_j\}$$

as a <u>level</u> trivial cofibration followed by a fibration

$$\{E_j\} \xrightarrow{\{i'_j\}} \{E'_j\} \xrightarrow{\{p'_j\}} \{B'_j\}.$$

Let $\{F'_j\}$ be the (<u>levelwise</u>) fibre of $\{p_j\}$. We obtain the following commutative diagram in $C_*^{\,J}$.

(3.4.18)
$$\begin{array}{ccccc}
\{F_j\} & \longrightarrow & \{E_j\} & \xrightarrow{\{p_j\}} & \{B_j\} \\
\downarrow{\scriptstyle \{i''_j\}} & & \downarrow{\scriptstyle \{i'_j\}} & & \downarrow{\scriptstyle \{i_j\}} \\
\{F'_j\} & \longrightarrow & \{E'_j\} & \xrightarrow{\{p'_j\}} & \{B'_j\}
\end{array}$$

In diagram (3.4.18) $\{i''_j\}$ is the restriction of $\{i'_j\}$. Regard diagram (3.4.18) as an inverse system made up of the diagrams

(3.4.19)
$$\begin{array}{ccccc}
F_j & \longrightarrow & E_j & \longrightarrow & B_j \\
\downarrow{\scriptstyle i''_j} & & \downarrow{\scriptstyle i'_j} & & \downarrow{\scriptstyle i_j} \\
F'_j & \longrightarrow & E'_j & \longrightarrow & B'_j
\end{array}$$

in C_*. Because p_j and p'_j are fibrations (a fibration in $C_*^{\,J}$ is a levelwise fibration, see Proposition (3.2.16), although the converse assertion need not hold), and i_j and i'_j are weak equivalences, the maps i''_j are weak equivalences. Hence the diagrams

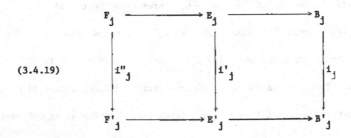

$$\{F_k\} \longrightarrow \{E_j\} \longrightarrow \{B_j\} ,$$

$$\{F'_j\} \longrightarrow \{E'_j\} \longrightarrow \{B'_j\}$$

are isomorphic over $\mathrm{Ho}(\mathrm{pro}-C_*)$. The conclusion follows. □

(3.4.20) Remarks. Diagrams (3.4.19) may be extended to maps of short fibration sequences in $\mathrm{Ho}(C_*)$.

(3.4.21)
$$\begin{array}{ccccccccc} F_j & \longrightarrow & E_j & \longrightarrow & B_j & & \Omega B_j \times F_j & \longrightarrow & F_j \\ \downarrow & & \downarrow & & \downarrow & & \downarrow & & \downarrow \\ F'_j & \longrightarrow & E'_j & \longrightarrow & B'_j, & & \Omega B'_j \times F'_j & \longrightarrow & F'_j . \end{array}$$

However diagrams (3.4.21) do not yield the required action map

$$\Omega\{B'_j\} \times \{F'_j\} \longrightarrow \{F'_j\}$$

in $\mathrm{Ho}(\mathrm{pro}-C_*)$ (not $\mathrm{pro}-\mathrm{Ho}(C_*)$), without suitable coherency conditions.

(3.4.22) Remarks. As in Quillen's discussion [Q-1, §1.3], we may obtain Toda brackets and similar constructions in $\mathrm{Ho}(\mathrm{pro}-C_*)$. Because we have no applications so far of these constructions, we omit the details.

§3.5. Simplicial Model Structures.

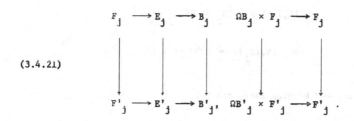

In this section we shall prove that a simplicial closed model structure (satisfying condition N of §2.3) on C induces such a structure on pro-C. These results can be readily extended to pointed categories; details are similar and omitted.

For a finite simplicial set K and X, Y, in C, let $X \otimes K$ and $HOM(X,Y)$ denote the "tensor product" and "function space" constructions in C (recall that $HOM(X,Y)$ is a simplicial set), and let $HOM(K,X)$ denote the "function space" connecting SS and C; see §2.4. Let $\{X_j\}, \{Y_k\} \in C$.

(3.5.1) Definition. Let

$$\{X_j\} \otimes K = \{X_j \otimes K\}, \quad \text{and}$$

$$HOM(K,\{X_j\}) = \{HOM(K,X_j)\}$$

together with the induced bonding maps, and let

$$HOM(\{X_j\}, \{Y_k\}) = \lim_k \operatorname{colim}_j \{HOM(X_j,Y_k)\}.$$

These constructions extend to functors $\otimes$: pro-$C \times SS \longrightarrow$ pro-C, HOM: (finite simplicial sets)$^{op} \times$ pro-$C \longrightarrow$ pro-C, and HOM: (pro-C)$^{op} \times$ pro-$C \longrightarrow SS$, respectively. Axiom SMO, and the following propositions are immediate consequences.

(3.5.2) Proposition. For $\{X_j\}$ and $\{Y_k\}$ in pro-C, the set of 0-simplices of $HOM(\{X_j\}, \{Y_k\})$,

$$HOM(\{X_j\}, \{Y_k\})_0 \cong \text{pro-}C(\{X_j\}, \{Y_k\}),$$

naturally in $\{X_j\}$ and $\{Y_k\}$.

(3.5.3) <u>Theorem</u> (Exponential Law). Let K be a finite simplicial set and let $\{X_j\}, \{Y_k\} \in$ pro-C. Then

$$HOM(\{X_j\} \otimes K, \{Y_k\}) \cong HOM(K, HOM(\{X_j\}, \{Y_k\})),$$

naturally in all variables. (HOM is used to denote the "function space" construction in both pro-C and SS.)

Proof. Because

$$\begin{aligned} HOM(\{X_j\} \otimes K, \{Y_k\}) &= HOM(\{X_j \otimes K\}, \{Y_k\}) \\ &= \lim_k\{colim_j\{HOM(X_j \otimes K, Y_k)\}\} \\ &= \lim_k \{HOM(\{X_j\} \otimes K, Y_k)\}, \end{aligned}$$

and $HOM(K,?): SS \longrightarrow SS$ preserves limits, we may reduce the general case to the case where $\{Y_k\}$ is an object Y of C.

In this case,

$$\begin{aligned} HOM(\{X_j\} \otimes K, Y) &= colim_j\{HOM(X_j \otimes K, Y)\} \\ &= colim_j\{HOM(K, HOM(X_j, Y))\}. \end{aligned}$$

Because K has finitely many non-degenerate simplices, and the indexing category $\{j\}$ is filtering,

$$\begin{aligned} colim_j\{HOM(K, HOM(X_j, Y))\} &= HOM(K, colim_j\{HOM(X_j, Y)\}) \\ &= HOM(K, HOM(\{X_j\}, Y)), \end{aligned}$$

as required. Naturality follows easily. □

(3.5.4) <u>Corollary</u>. With $K, \{X_j\}$, and $\{Y_k\}$ as above,

$$\text{pro-}C(\{X_j\} \otimes K, \{Y_k\}) = SS(K, \text{HOM}(\{X_j\}, \{Y_k\})),$$

naturally in all variables. □

(3.5.5) <u>Remarks</u>.

a) The corresponding assertion for HOM (K,X) (Definition (2.4.2)) is proven similarly. Details are omitted.

b) The above results fail for infinite K; to construct counter-examples, use the fact that $\text{HOM}(K,-): SS \longrightarrow SS$ does <u>not</u> commute with colimits for infinite K. Similarly the function space $\text{MAP}(K,-): \text{Top} \longrightarrow \text{Top}$ does <u>not</u> commute with expanding colimits for non-compact K.

(3.5.6) <u>Theorem</u> (Verification of Axiom SM7). Let $i: \{A_j\} \longrightarrow \{X_k\}$ be a cofibration in pro-C, and let $p: \{Y_\ell\} \longrightarrow \{B_m\}$ be a fibration in pro-C. Then:

a) The induced map

(3.5.5)

$$\text{HOM}(\{X_k\}, \{Y_\ell\}) \xrightarrow{q} \text{HOM}(\{A_j\}, \{Y_\ell\}) \times_{\text{HOM}(\{A_j\}, \{B_m\})} \text{HOM}(\{X_k\}, \{B_m\})$$

is a fibration in SS (i.e., a Kan fibration);

b) If either i or p is also a weak equivalence, then the map q above is also a weak equivalence.

<u>Proof</u>.

a) Consider a solid-arrow commutative diagram in SS of the form

(3.5.6)

$$\begin{array}{ccc}
V^{n,k} & \xrightarrow{\quad f \quad} & \mathrm{HOM}\ (\{K_k\},\{Y_\ell\}) \\
\Big\downarrow & & \Big\downarrow \\
\Delta^n & \xrightarrow{g} & \mathrm{HOM}\ (\{A_j\},\{Y_\ell\}) \times_{\mathrm{HOM}\ (\{A_j\},\{B_m\})} \mathrm{HOM}\ (\{X_k\},\{B_m\})
\end{array}$$

in which $V^{n,k}$ is obtained from $\partial\Delta^n$, the boundary of Δ^n, by deleting the kth face. The maps g and q correspond respectively to pairs of maps

(3.5.7)
$$g':\Delta^n \longrightarrow \mathrm{HOM}\ (\{A_j\},\{Y_\ell\}),$$
$$g'':\Delta^n \longrightarrow \mathrm{HOM}\ (\{X_k\},\{B_m\});$$

$$q':\mathrm{HOM}\ (\{X_k\},\{Y_\ell\}) \longrightarrow \mathrm{HOM}\ (\{A_j\},\{Y_\ell\}),$$
$$q'':\mathrm{HOM}\ (\{X_k\},\{Y_\ell\}) \longrightarrow \mathrm{HOM}\ (\{X_k\},\{B_m\});$$

such that the appropriate composite maps into $\mathrm{HOM}\ (\{A_j\},\{B_m\})$ are equal. Applying the exponential law (Theorem (3.5.3)) to f, g', g", q', and q", and assembling the induced maps with the above coherence data (Diagram (3.5.6)), we obtain a commutative solid-arrow diagram

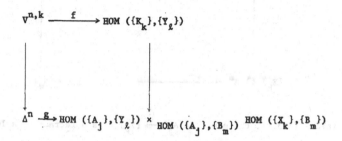

(3.5.8)

$$\begin{array}{ccc}
\{A_j\} \otimes \Delta^n \cup_{\{A_j\} \otimes V^{n,k}} \{X_k\} \otimes V^{n,k} & \longrightarrow & \{Y_\ell\} \\
\downarrow & & \downarrow \\
\{X_k\} \otimes \Delta^n & \longrightarrow & \{B_m\}.
\end{array}$$

By Definitions (3.3.1), the map $i:\{A_j\} \longrightarrow \{X_k\}$ is a retract of a levelwise cofibration $\{i'_r\}:\{A'_r\} \longrightarrow \{X'_r\}$. By Axiom SM7 for C, (see §2.4) the induced maps

(3.5.9) $$A'_r \otimes \Delta^n \cup_{A'_r \otimes V^{n,k}} X'_r \otimes V^{n,k} \longrightarrow X'_r \otimes \Delta^n$$

are trivial cofibrations; hence the induced map

(3.5.10)

$$\{i'_r\}_*:\{A'_r\} \otimes \Delta^n \cup_{\{A'_r\} \otimes V^{n,k}} \{X'_r\} \otimes V^{n,k} \longrightarrow \{X'_r\} \otimes \Delta^n$$

is a trivial cofibration. Because i_* is a retract of $\{i'\}_*$ by construction, i_* is also a trivial cofibration; hence Axiom M1 for pro-C yields the filler h' in diagram (3.5.8). Applying the exponential law (Theorem 3.5.3)) to h' yields a map $h:\Delta^n \longrightarrow \mathrm{HOM}(\{X_k\},\{Y_\ell\})$. By construction ((3.5.7) - (3.5.10)), the map h makes diagram (3.5.6) commute. Thus q is a (Kan) fibration, as required.

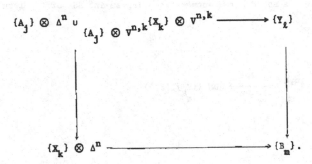

b) If the map i is a trivial cofibration, then i induces a trivial cofibration

(3.5.11) $$i_{\#}:\{A_j\} \otimes \Delta^n \cup_{\{A_j\} \otimes \partial\Delta^n} \{X_k\} \otimes \partial\Delta^n \longrightarrow \{X_k\} \otimes \Delta^n$$

by analogues of (3.5.9) with $V^{n,k}$ replaced by $\partial\Delta^n$; hence a filler in the analogue of diagram (3.5.6) with the same replacement. Thus q is a trivial (Kan) fibration, as required.

Finally, suppose that p is a trivial fibration. The cofibration $i_{\#}$ (3.5.11) induced by i has the left-lifting-property with respect to p, so the required fillers in suitable analogues of diagrams (3.5.8) and (3.5.6) (see above) exist. Thus q is a trivial (Kan) fibration, as required. □

Theorems (3.5.3) and (3.5.6), together with the earlier proof that pro - C is a closed model category (§§3.2 - 3.3), imply that pro - C is a simplicial closed model category.

§3.6. Pairs.

We shall use the Bousfield-Kan [B -K] model structure on Maps(pro - C). More precisely, Maps(pro - C) is the category whose objects are maps

$$A \longrightarrow X$$

in pro - C, and whose morphisms are commutative squares. A map

(3.6.1)
$$\begin{array}{ccc} A & \xrightarrow{f} & B \\ \downarrow & & \downarrow \\ X & \xrightarrow[g]{} & Y \end{array}$$

in Maps(pro - C) is a weak equivalence if g and f are weak equivalences, a fibration if g and f are fibrations, and a cofibration if the appropriate lifting property is satisfied. Explicitly, (g,f) is a cofibration if f and the induced map

$$X \coprod_A B \longrightarrow Y$$

are cofibrations in pro - C ($X \coprod_A B$ is the pushout obtained from most of diagram (3.6.1)).

(3.6.2) <u>Definition</u>. Let (C, pro - C) be the full subcategory of Maps(pro - C) consisting of maps $A \longrightarrow X$ with X in C.

It is easy to prove (as in §3.2):

(3.6.3) <u>Theorem</u>. The closed model structure on pro - C induces natural closed model structures on Maps(pro - C), (C, pro - C), and (C, tow - C). □

It is convenient to represent (C, tow - C) as follows. Objects are towers $X = (X_0 \longleftarrow X_1 \longleftarrow \cdots)$; a morphism consists of a map $f: X \longrightarrow Y$ in tow - C together with a compatible map $f_0: X_0 \longrightarrow Y_0$ in C. Alternatively, a map $f:\{X_m\} \longrightarrow \{Y_n\}$ consists of a cofinal subtower $\{X_{m_n}\} \subset \{X_m\}$ with $X_{m_0} = X_0$ and a level map $\{X_{m_n}\} \longrightarrow \{Y_n\}$.

3.7. Geometric Models.

We shall discuss geometric models of Ho(Top, tow - Top) and Ho(tow - Top) using filtered spaces and a telescope construction. The geometric model of Ho(Top, tow - Top) will be used in proper homotopy theory in §6. The model of Ho(tow - Top) was used in shape theory by the first author and R. Geoghegan (unpublished).

(3.7.1) Definitions. A filtered space X consists of an underlying space X together with a sequence of closed subspaces $X = X_0 \supset X_1 \supset X_2 \supset \cdots$, with each $X_n \subset \operatorname{int} X_{n+1}$. A filtered map $f: X \longrightarrow Y$ (of filtered spaces) is a continuous map such that for each number $n \geq 0$ there is a number $m \geq 0$ with $f(X_m) \subset Y_n$. A filtered space X induces a natural filtration on its cylinder $X \times [0,1]$; this yields a natural notion of filtered homotopy of filtered maps. There results a filtered category Filt and its associated filtered homotopy category (quotient under the relation of filtered homotopy) Ho(Filt).

(3.7.2) Definition. The telescope of a tower $X = \{X_n\}$ is the space

$$\operatorname{Tel}(X) = X_0 \times 0 \cup_{\text{bond}} X_1 \times [0,1] \cup_{\text{bond}} X_2 \times [1,2] \cup_{\text{bond}} \cdots$$

shown below.

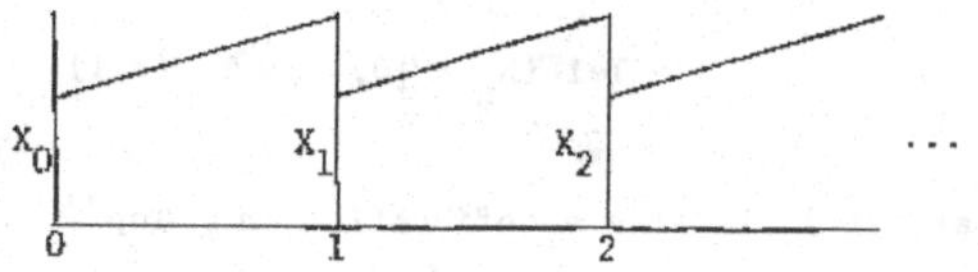

Tel (X) is filtered by setting

$$\mathrm{Tel}\ (X)_n = X_n \times n \cup_{\mathrm{bond}} X_{n+1} \times [n, n+1] \cup_{\mathrm{bond}} \cdots .$$

This construction extends to a functor

$$\mathrm{Tel}\colon\ \mathrm{Top}^N \longrightarrow \mathrm{Filt}.$$

Further, Tel takes a "$\times [0,1]$ - homotopy" in Top^N into a filtered homotopy.

(3.7.3) Definition. The category Tel of telescopes is the full subcategory of Filt consisting of telescopes.

This should cause no confusion. Observe that the functor Tel factors through the telescope category Tel; also that $\mathrm{Tel}\ \{X_n\} \times [0,1] = \mathrm{Tel}\ \{X_n \times [0,1]\}$ so that cylinders may be formed within Tel. We therefore let Ho(Tel) be the full subcategory of telescopes in Ho(Filt).

(3.7.4) Proposition. Let $\{f_n\}:\{X_n\} \longrightarrow \{Y_n\}$ be a weak equivalence in Top^N. Then $\mathrm{Tel}\ \{f_n\}$ is invertible in Ho(Tel).

Proof. Observe that $\mathrm{Tel}\ \{X_n\}$ is a strong deformation retract in Tel of the mapping cylinder

$$\begin{aligned}\mathrm{Map}\ (\mathrm{Tel}\ \{f_n\}) &= \mathrm{Tel}\ \{\mathrm{Map}\ (f_n)\} \\ &= \mathrm{Tel}\ \{X_n \times [0,1] \cup Y_n/(x,1) \sim f(x)\}.\end{aligned}$$

We thus assume that $\{f_n\}$ is also a cofibration in Top^N.

For each n choose a retraction $r_n : Y_n \longrightarrow X_n$, and a homotopy $H_n : Y_n \times I \longrightarrow Y_n$ with $H_{n|0} = id$ and $H_{n|1} = f_n r_n$. <u>Caution</u>: in general

$$bond \circ r_{n+1} \neq r_n \circ bond, \quad \text{and}$$

$$bond \circ H_{n+1} \neq H_n \circ bond.$$

However,

$$(3.7.5) \qquad \begin{aligned} bond \circ r_{n+1} \circ f_{n+1} &= bond \\ &= r_n \circ f_n \ \ bond \\ &= r_n \circ bond \circ f_{n+1}, \quad \text{and} \end{aligned}$$

$$(3.7.6) \qquad bond \circ H_{n+1} \circ (f_{n+1} \times id) = H_n \circ bond \circ (f_{n+1} \times id) : \\ X_{n+1} \times [0,1] \longrightarrow Y_n .$$

We shall now use (3.7.5) to define a filtered map $g: \mathrm{Tel}\ \{Y_n\} \longrightarrow \mathrm{Tel}\ \{X_n\}$, which will be shown to be a filtered-homotopy-inverse to $\mathrm{Tel}\ \{f_n\}$. Because the map f_{n+1} is a trivial cofibration, $Y_{n+1} \times 0 \cup X_{n+1} \times [0,1] \cup Y_{n+1} \times 1$ is a strong deformation retract of $Y_{n+1} \times [0,1]$ for $n = 0$. Let ρ_{n+1} be the retraction. The composite map, to be denoted K_{n+1},

$$(3.7.7) \qquad Y_{n+1} \times [0,1] \xrightarrow{\rho_{n+1}} Y_{n+1} \times 0 \cup X_n \times [0,1] \cup Y_{n+1} \times 1$$

$$\xrightarrow{r_n \circ \text{bond} \,\cup\, r_n \circ \text{bond} \circ f_{n+1} \,\cup\, \text{bond} \circ r_{n+1}} X_n$$

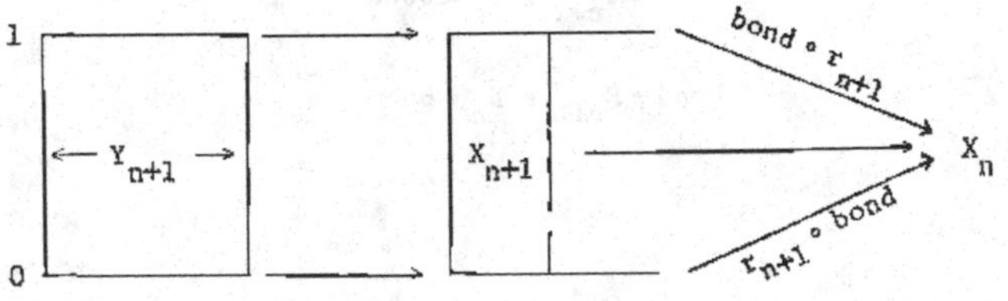

(see (3.7.5)) yields a homotopy from $r_n \circ \text{bond}$ to $\text{bond} \circ r_{n+1}$. Now define a map $g:\text{Tel}\,\{Y_n\} \longrightarrow \text{Tel}\,\{X_n\}$ as follows: g maps $Y_{n+1} \times [n + \frac{1}{2}, n+1]$ to $X_{n+1} \times [n, n+1]$ according to the formula

$$(3.7.8) \qquad g(y,t) = (r_{n+1}(y),\ 2t - n - 1);$$

g maps $Y_{n+1} \times [n, n+\frac{1}{2}]$ to $X_n \times n$ according to the formula

$$(3.7.9) \qquad g(y,t) = (K_{n+1}(y,\ 2t - 2n),\ n);$$

and g maps $Y_n \times n$ to $X_n \times n$ according to the formula

$$(3.7.10) \qquad g(y,t) = (r_n(y),\ n).$$

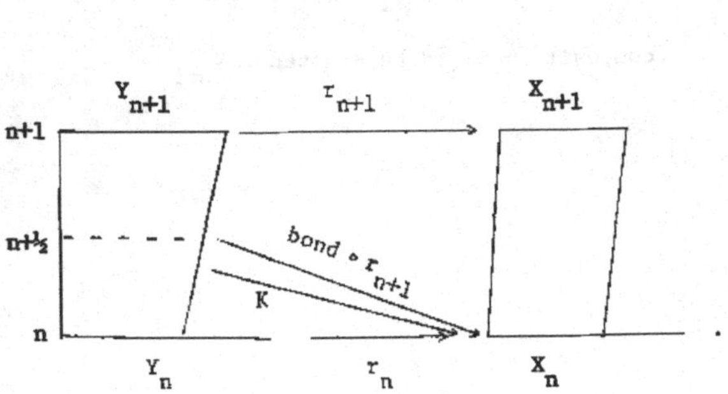

Then g is a filtered map.

<u>Claim 1</u>: The maps gf and $id_{Tel\ \{X_n\}}$ are filtered-homotopic. To check this, first observe that $K_{n+1}|X_{n+1} \times [0,1]$ is the projection onto X_{n+1} by (3.7.5). Thus

$$(3.7.11) \qquad gf(x,t) = \begin{cases} (x,\ 2t-n-1), & n+\tfrac{1}{2} \leq t \leq n+1, \\ (x,n), & n \leq t \leq n+\tfrac{1}{2}, \end{cases}$$

by (3.7.8) - (3.7.10). Claim 1 follows easily.

<u>Claim 2</u>: The maps fg and $id_{Tel\ \{Y_n\}}$ are filtered-homotopic. We shall use (3.6.6) to imitate the construction of g and obtain the required homotopy. As above, $X_{n+1} \times [0,1] \times [0,1]' \cup Y_{n+1} \times \partial([0,1] \times [0,1]')$ is a strong deformation retract of $Y_{n+1} \times [0,1] \times [0,1]'$. ($[0,1]'$ is just a second unit interval.) We may therefore define a homotopy Γ_{n+1} from K_{n+1} to bond rel $X_{n+1} \times [0,1] \times [0,1]' \cup Y_{n+1} \times \partial([0,1] \times [0,1]')$ schematically,

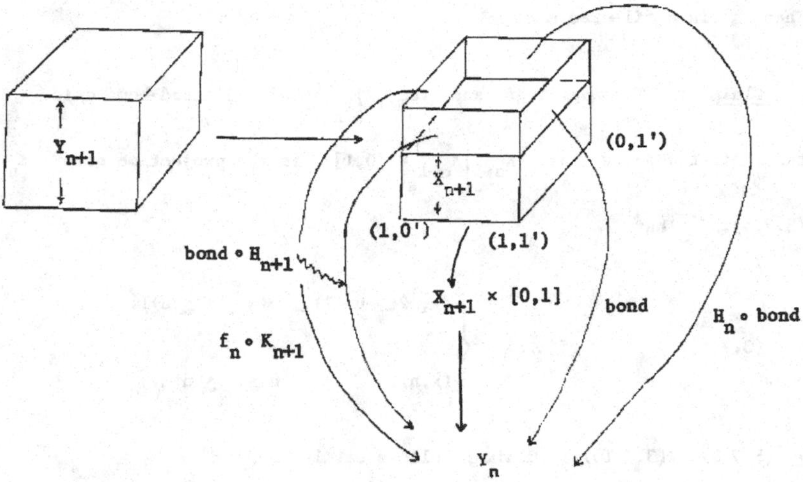

Gluing these maps together yields a filtered homotopy

$$\Gamma:\text{Tel }\{Y_n\} \times [0,1]' \longrightarrow \text{Tel }\{Y_n \times [0,1]'\} \longrightarrow \text{Tel }\{Y_n\}$$

from fg to a map $h:\text{Tel }\{Y_n\} \longrightarrow \text{Tel }\{Y_n\}$ which changes only vertical coordinates, and moves those coordinates at most ½ unit as in (3.7.11). Details are analogous to those in Claim 1 and are omitted. Claim 2 follows. □

(3.7.12) <u>Corollary</u>. The functor $\text{Tel}:\text{Top}^N \longrightarrow \text{Ho(Tel)}$ factors through $\text{Ho}(\text{Top}^N)$ to induce

$$\text{Tel}:\text{Ho}(\text{Top}^N) \longrightarrow \text{Ho(Tel)}.$$

<u>Proof</u>. By the model structure, each map in $\text{Ho}(\text{Top}^N)$ may be represented by a diagram

$$\begin{array}{ccc} \{X_n\} & & \{Y_n\} \\ & \searrow & \swarrow {\scriptstyle \{j_n\}} \\ & \{Y'_n\} & \end{array}$$

where $\{Y'_n\}$ is fibrant in Top^N (i.e., a tower of fibrations) and $\{j_n\}$ is a (levelwise) weak equivalence. (Note: all objects in Top^N are cofibrant). Any two such maps are homotopic with respect to the cylinder $-\times[0,1]$. Further, a weak equivalence between fibrant objects in Top^N admits a homotopy inverse with respect to $-\times[0,1]$. The conclusion follows by Proposition (3.7.4). □

We shall now extend Tel to Ho(Top, tow - Top).

(3.7.13) <u>Proposition</u>. Suppose that $\{X_{n_k}\}$ is a cofinal subtower of $\{X_n\}$ with $X_{n_0} = x_0$. Then there is a natural equivalence $\mathrm{Tel}\,\{X_n\} \longrightarrow \mathrm{Tel}\,\{X_{n_k}\}$ in Ho(Tel).

<u>Proof</u>. The required map $\mathrm{Tel}\,\{X_n\} \longrightarrow \mathrm{Tel}\,\{X_{n_k}\}$ is defined by mapping $X_{n_k} \times n_k$ to $X_{n_k} \times k$ under the identity and extending to $\mathrm{Tel}\,\{X_n\}$ as illustrated.

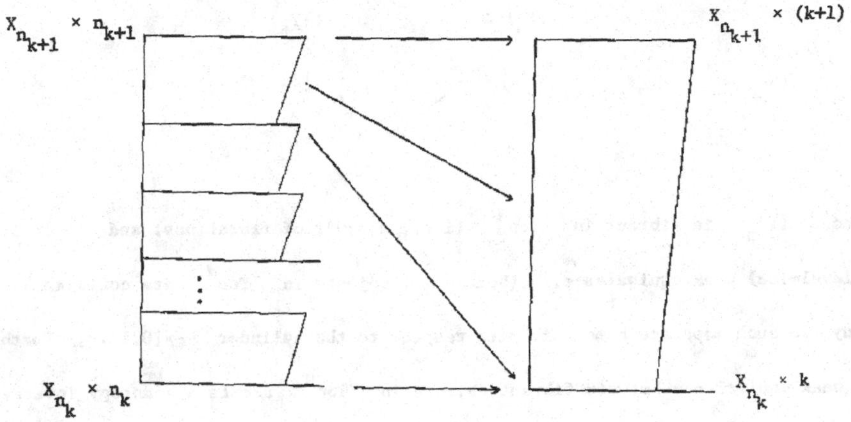

This map is easily seen to be a filtered homotopy equivalence. □

(3.7.14) Proposition.

a) The functor $\mathrm{Tel}:\mathrm{Top}^{N} \longrightarrow \mathrm{Ho}(\mathrm{Tel})$ factors through (Top, tow-Top).

b) This functor factors through Ho(Top, two-Top), to induce

$$\mathrm{Tel}:\mathrm{Ho}(\mathrm{Top},\ \mathrm{tow\text{-}Top}) \longrightarrow \mathrm{Ho}(\mathrm{Tel}).$$

Proof. Part (a) follows from Proposition (3.7.13) because each map in (Top, tow-Top) may be represented by a diagram

$$\{X_n\} \supset \{X_{n_k}\} \longrightarrow \{Y_k\}$$

where $\{X_{n_k}\}$ is cofinal in $\{X_n\}$, $\{X_{n_0}\} = X_0$, and $\{f_k\}$ is a level map (i.e., a map in Top^{N}).

Part (b) follows from Part (a) and the observation that Y_n is fibrant in

(Top, tow-Top) if $\{Y_n\}$ is a tower of fibrations by using the proof of Corollary (3.7.12). □

We are ready to show that Tel:Ho(Top, tow-Top) ⟶ Ho(Tel) yields an equivalence of categories.

(3.7.15) Definition. The end of a filtered space X is the tower $\varepsilon(X) = \{X_n\}$ in (Top, tow-Top). Then ε extends to a functor

$$\varepsilon: \text{Tel} \longrightarrow (\text{Top, tow-Top}).$$

The definition of Ho(Tel) implies the following.

(3.6.16) Proposition. ε induces a functor

$$\varepsilon: \text{Ho(Tel)} \longrightarrow \text{Ho(Top, tow-Top)}.$$

(3.7.17) Definitions. Let $X = \{X_n\} \in \text{Top}^N$. Let $p: \varepsilon \circ \text{Tel}\ (X) \longrightarrow X$ be the map on Top^N given by letting $p_n | X_k \times [k-1, k]$ be the composite

$$X_k \times [k-1, k] \longrightarrow X_k \xrightarrow{\text{bond} \circ \cdots \circ \text{bond}} X_n$$

for $k > n$. Let

$$q = \text{Tel}\ (p): (\text{Tel} \circ \varepsilon)(\text{Tel}\ (X)) \longrightarrow \text{Tel}\ (X).$$

(3.7.18) Proposition.

a) The maps p are natural weak equivalences in Ho(Top, tow-Top).

b) The maps q are natural weak equivalence in Ho(Tel).

Proof. Part (a) follows immediately from the definitions of Tel, ε, and p (note that each p_n is an equivalence in Ho(Top)). Part (b) follows from Part (a), the definition of the telescope category and Proposition (3.7.14b). □

Propositions (3.7.14b), (3.7.16), and (3.7.18) imply the following:

(3.7.19) Theorem. The categories Ho(Top, tow-Top) and Ho(Tel) are naturally equivalent. □

A geometric model for Ho(tow-Top) may be obtained as follows. We may assume that all towers $X = \{X_n\}$ satisfy $X_0 = *$. This embeds tow-Top in (Top, tow-Top) (regard X as (*,X)), and gives rise to a full subcategory of Tel, ConTel (contractible telescopes) consisting of those telescopes Tel $(* = X_0 \longleftarrow X_1 \longleftarrow \cdots)$.

(3.7.20) Corollary. The categories Ho(tow-Top) and Ho(ConTel) are naturally equivalent.

(3.7.21) Remarks. In R. Vogt's [Vogt-1] approach to the homotopy theory of categories of diagrams, Tel $\{X_n\}$ is the homotopy colimit of the diagram $\{X_0 \longleftarrow X_1 \longleftarrow \cdots\}$. In this setting our homotopy category Ho(Tel) represents "coherent pro-homotopy," that is, a version of pro-homotopy where maps and homotopies are required to satisfy various coherency conditions. The development of coherent pro-homotopy theory (Vogt only works with level maps) is an interesting problem whose solution should have applications to proper homotopy theory (see §6), and shape theory, especially in alternative proofs of the Chapman [Chap-1] complement theorem (see §6).

§3.8. Inj - spaces.

The above theory of pro -spaces admits a straight-forward dualization to inj -spaces (direct systems). We shall sketch this theory without proofs.

Let C be a closed model category which satisfies Condition N of §2.3. Let J (= {j}) be a cofinite strongly directed set. Then $C^{J^{op}}$ is the category of direct systems over C indexed by J. The closed model structure on the level category $SS^{J^{op}}$ is due to Bousfield and Kan [B - K, p. 314].

(3.8.1) Definitions [B -K, p. 314] (dual to (3.2.1)). A map $f:X \to Y$ ($= \{f_j:X_j \longrightarrow Y_j\}$) in $C^{J^{op}}$ is a fibration (resp., weak equivalence) if for all j in J, the maps f_j are fibrations (resp., weak equivalences) in C.

A map f in $C^{J^{op}}$ is a cofibration if it has the left-lifting property with respect to all maps p which are both fibrations and weak equivalences.

(3.8.2) Theorem (dual to (3.2.2)). $C^{J^{op}}$, together with the above structure is a closed model category.

(3.8.3) Proposition (dual to (3.2.3)). A map $f:X \to Y$ in $C^{J^{op}}$ is a cofibration if for each j in J the induced map q_j in the diagram

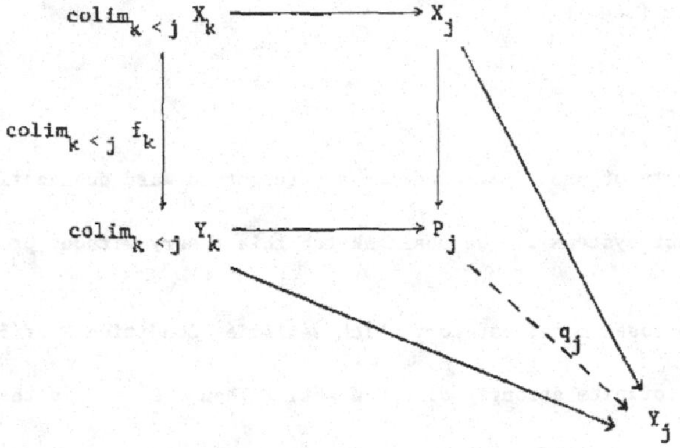

(P_j is the pushout) is a cofibration.

This result, in the case $J = N$, is the usual definition of cofibration of CW spectra (see [Vogt - 2]).

(3.8.4) Theorem (dual to (3.2.4)). The constant diagram functor $C \longrightarrow C^{J^{op}}$ preserves cofibrations, fibrations, and weak equivalences. The direct limit functor $\text{colim}: C^{J^{op}} \longrightarrow C$ preserves cofibrations and trivial cofibrations (maps which are both cofibrations and weak equivalences).

We shall now discuss the homotopy theory of inj - C.

(3.8.5) Definitions (dual to (3.3.1)). A map f in inj - C is called a strong cofibration if f is the image in inj - C of a (level) cofibration $\{f_j\}$ in some $C^{J^{op}}$ where J is a cofinite strongly directed set. Strong fibrations, strong trivial cofibrations, and strong trivial fibrations are defined similarly.

A map in inj-C is a <u>cofibration</u> if it is the retract in Maps(inj-C) of a strong cofibration. Fibrations, trivial cofibrations, and trivial fibrations in inj-C are defined similarly. A map f in inj-C is a weak equivalence if f = pi where p is a trivial fibration and i is a trivial cofibration.

(3.8.6) <u>Theorem</u> (dual to (3.3.3)). inj-C, together with the above structure, is a closed model category.

(3.8.7) <u>Theorem</u> (dual to (3.3.4)). The constant diagram functor $C \longrightarrow \text{inj-}C$ preserves cofibrations, fibrations, and weak equivalences. The direct limit functor colim: $\text{inj-}C \longrightarrow C$ preserves cofibrations and trivial cofibrations.

The results of §§3.4 - 3.7 may also be dualized to inj-C. We shall simply note for later reference that the full subcategory of <u>direct towers</u> (direct systems of spaces indexed by the natural numbers) contained in Ho(inj-C) admits a geometric model along the lines of §3.7.

(3.8.8) Definition (Milnor [Mil-3], dual to (3.7.2)). The telescope of a direct tower $X = \{X_n\}$ is the space

$$\text{Dir tel}(X) = X_0 \times [0,1] \cup_{\text{bond}} X_1 \times [1,2] \cup_{\text{bond}} \cdots$$

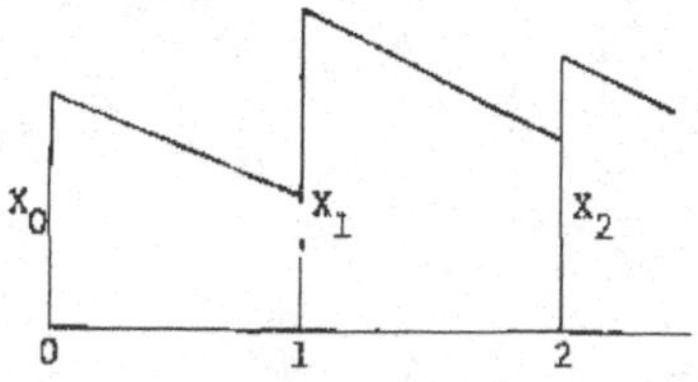

Dir tel (X) admits a natural <u>increasing</u> filtration (we call this a cofiltration). □

We shall now consider those direct towers $X = \{X_n\}$ with $X_0 = *$, and contractible direct telescopes (again, $X_0 = *$). Any direct tower is equivalent in inj -C to a direct tower $X = \{X_n\}$ with $X_0 = *$.

(3.8.9) <u>Theorem</u> (dual to (3.7.20)). The direct telescope functor induces an equivalence of homotopy categories from Ho(inj - C) to the homotopy category of contractible direct telescopes and cofiltered maps.

4. THE HOMOTOPY INVERSE LIMIT AND ITS APPLICATIONS TO HOMOLOGICAL ALGEBRA

§4.1. Introduction.

In this chapter we shall define a homotopy inverse limit functor $\text{holim}: \text{Ho}(\text{pro}-C) \longrightarrow \text{Ho}(C)$ adjoint to the inclusion functor $\text{Ho}(C) \longrightarrow \text{Ho}(\text{pro}-C)$ for suitable closed model categories C (§4.2) and obtain applications to the study of the derived functors $\lim^s$ of the inverse limit. Bousfield and Kan [B -K; Chapter XI] gave a less intuitive construction of a (somewhat different) homotopy inverse limit on $\text{pro}-SS^J$ (see §§4.2, 4.9), and suggested the study of homotopy inverse limits on other pro-categories. §4.3 is an appendix to §4.2.

In §4.4 we survey some of the results of other authors, notably Z. Z. Yeh, J.-E. Roos, J.-L. Verdier, C. Y. Jensen, and B. Osofsky, on $\lim^s$ as background for our own work. Our results are stated in §4.5 and proved in the following sections: §4.6, Algebraic description of $\lim^s$; §4.7, Topological description of $\lim^s$; §4.8, Vanishing theorems for $\lim^s$.

In this chapter, C shall denote a closed model category which satisfies Condition N (§2.3) and <u>admits</u> <u>arbitrary</u> (inverse) <u>limits</u>. In particular, we shall discuss $C = SS$, SSG (simplicial groups) and $SSAG$ (simplicial abelian groups).

§4.2 The homotopy inverse limit.

In this section we shall define a homotopy inverse limit functor

$$\text{holim}: \text{Ho(pro-C)} \longrightarrow \text{Ho(C)}$$

adjoint to the inclusion $\text{Ho(C)} \longrightarrow \text{Ho(pro-C)}$. That is, for X in C and $\{Y_j\}$ in pro-C,

(4.2.1) $$\text{Ho(pro-C)}(X, \{Y_j\}) \cong \text{Ho(C)}(X, \text{holim}\{Y_j\}).$$

We begin by considering formula (4.2.1) in the case that X is cofibrant in C, and hence in pro-C, and $\{Y_j\}$ is fibrant in pro-C.

(4.2.2) Proposition. The inverse limit functor $\lim:(\text{pro-C})_{cf} \longrightarrow C$ induces a functor on the homotopy categories

$$\lim: \text{Ho}((\text{pro-C})_{cf}) \longrightarrow \text{Ho(C)}.$$

Proof. To define lim on maps in $\text{Ho}((\text{pro-C})_{cf})$, recall that for $\{X_j\}, \{Y_k\} \in (\text{pro-C})_{cf}$,

(4.2.3) $$\text{Ho}((\text{pro-C})_{cf})(\{X_j\},\{Y_k\}) = [\{X_j\},\{Y_k\}],$$

the set of homotopy classes of maps from $\{X_j\}$ to $\{Y_k\}$ with respect to a cylinder $\{X_j\} \otimes [0,1]$. Let $f,g:\{X_j\} \longrightarrow \{Y_k\}$ be homotopic maps in pro-C, and let $H:\{X_j\} \otimes [0,1] \longrightarrow Y$ be a homotopy with $Hi_0 = f$ and $Hi_1 = g$. Applying the inverse limit to the diagram

$$(4.2.4a) \qquad \begin{array}{l} \{X_j\} \searrow^{i_0} \\ \qquad\qquad \{X_j\} \otimes [0,1] \xrightarrow{H} \{Y_k\} \\ \{X_j\} \nearrow_{i_1} \end{array}$$

yields the diagram

$$\begin{array}{l} \lim \{X_j\} \searrow^{\lim i_0} \\ \qquad\qquad \lim (\{X_j\} \otimes [0,1]) \xrightarrow{\lim H} \lim \{Y_k\}. \\ \lim \{X_j\} \nearrow_{\lim i_1} \end{array}$$

Now observe that the maps i_0 and i_1 in diagram (4.2.4a) are sections to the <u>trivial fibration</u> $p:\{X_j\} \otimes [0,1] \longrightarrow \{X_j\}$. Hence, $\lim i_0$ and $\lim i_1$ are sections to the <u>induced</u> (lim preserves trivial fibrations, see Theorem (3.3.4)) <u>trivial fibration</u> $\lim p: \lim (\{X_j\} \otimes [0,1]) \longrightarrow \lim \{X_j\}$. This implies that the maps $\lim i_0$ and $\lim i_1$ are weak equivalences. By diagram (4.2.5), the maps

$$\lim f = \lim H \circ \lim i_0$$

and

$$\lim g = \lim H \circ \lim i_1$$

become equivalent in $Ho(C)$. □

(4.2.5) <u>Proposition</u>. The inverse limit functor $\lim:(\text{pro}-C) \longrightarrow C$ induces a functor on the homotopy categories,

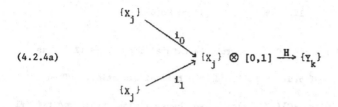

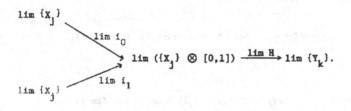

$$\lim\colon \ \mathrm{Ho}((\text{pro-}C)_f) \longrightarrow \mathrm{Ho}(C).$$

Proof. Let $\{X_j\} \in \text{pro-}C$. Factor the natural map $\phi \longrightarrow \{X_j\}$ as $\phi \longrightarrow \{X_k\} \longrightarrow \{X_j\}$; a cofibration followed by a trivial fibration. Then $\mathrm{Ho}(\text{pro-}C)(\{X_j\},-) = \mathrm{Ho}(\text{pro-}C)(\{X_k\},-)$; also, because lim preserves trivial fibrations, $\mathrm{Ho}(C)(\lim \{X_j\},-) = \mathrm{Ho}(C)(\lim \{X_k\},-)$. The conclusion now follows from Proposition (4.2.2). □

(4.2.6) Proposition. For X cofibrant in C and $\{Y_j\}$ fibrant in pro-C,

$$\mathrm{Ho}(\text{pro-}C)\ (X,\ \{Y_j\}) \cong \mathrm{Ho}(C)\ (X,\ \lim\ \{Y_j\}). \tag{4.2.7}$$

Proof. Formula (4.2.7) follows easily from adjointness of lim on pro-C,

$$\text{pro-}C\ (X,\ \{Y_j\}) = C\ (X,\ \lim\ \{Y_j\}),$$

and Proposition (4.2.2). □

We are now ready to define the homotopy inverse limit. By Quillen's theory of model categories (see §2.3), the homotopy theory of fibrant objects in pro-C, $\mathrm{Ho}((\text{pro-}C)_f)$ is equivalent to the homotopy theory of pro-C, $\mathrm{Ho}(\text{pro-}C)$. Following Quillen, for each object X in pro-C, factor the map $X \longrightarrow *$ as a trivial cofibration i_X followed by a fibration

$$X \xrightarrow{i_X} \mathrm{Ex}^{\infty}X \longrightarrow * \tag{4.2.8}$$

(If X is fibrant, choose $\mathrm{Ex}^{\infty}X = X$). This construction induces a functor

(4.2.9) $\qquad Ex^{\infty}: Ho(pro-C) \longrightarrow Ho((pro-C)_f);$

Ex^{∞} is defined on morphisms by applying Axiom M6 to obtain fillers in the diagrams

$$\begin{array}{ccc} X & \xrightarrow{\quad f \quad} & Y \\ {\scriptstyle i_X}\downarrow & & \downarrow{\scriptstyle i_Y} \\ Ex^{\infty}X & \xrightarrow{\; Ex^{\infty}f \;} & Ex^{\infty}Y \end{array} .$$

The homotopy class of $Ex^{\infty}f$ depends only on the homotopy class of f because

$$[Ex^{\infty}f] = [i_Y][f][i_X]^{-1}.$$

We therefore define the <u>homotopy inverse limit</u>, holim, to be the composite functor

(4.2.10)

$$holim \equiv \lim \circ Ex^{\infty} : Ho(pro-C) \longrightarrow Ho((pro-C)_f) \longrightarrow Ho(C).$$

(4.2.11) <u>Remarks</u>. If $X \in pro-C$ is fibrant, we may take $Ex^{\infty}X = X$ in (4.2.8). Hence, $holim\ X \cong \lim X$ on $(pro-C)_f$.

Propositions (4.2.5) and (4.2.6) immediately yield the following.

(4.2.12) <u>Theorem</u>. The functor $holim: Ho(pro-C) \longrightarrow Ho(C)$ is adjoint to the inclusion $Ho(C) \longrightarrow Ho(pro-C)$.

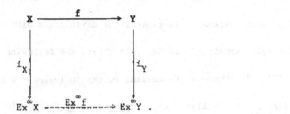

(4.2.13) Remarks. Bousfield and Kan [B -K, Chapter XI] defined a different "homotopy inverse limit" functor $\text{holim}_{B-K} : \text{Ho}(SS^J) \longrightarrow \text{Ho}(SS)$. They observed that for a tower of fibrations X, $\text{holim}_{B-K} X \sim \lim X$. But, for a tower of fibrations X, we may take $\text{Ex}^{\infty} X = X$; hence their definition is equivalent with ours on such systems. In general our definitions differ except on fibrant objects. For example, Bousfield and Kan only obtain the following analogue of Theorem (4.2.11): $R\ \text{holim}_{B-K}$ is adjoint to the inclusion $\text{Ho}(SS) \longrightarrow \text{Ho}(SS^J)$, where $R\ \text{holim}_{B-K}$ is Quillen's total right derived functor [Q -1, §1.4] of holim_{B-K}. In fact, for our holim, $\text{holim} = R\ \text{holim} = R\ \lim$, where lim is the ordinary inverse limit.

§4.3 Ex^{∞} on pro - SS.

In this section we shall describe an explicit Ex^{∞} functor from Ho(pro -SS) to Ho(pro -SS)_f, together with natural (in Ho(pro -SS)) trivial cofibrations $X \longrightarrow \text{Ex}^{\infty} X$ in pro -SS. Compare (4.2.8) - (4.2.9).

(4.3.1) Ex^{∞} on objects. Let $X \in$ pro - SS. First apply the Mardešić construction (Theorem (2.1.6)) which replaces X functorially by a naturally isomorphic object $MX = \{X_j\}$ indexed by a cofinite strongly directed set $J = \{j\}$. We thus need only define $\{Z_j\} = \text{Ex}^{\infty}\{X_j\}$ where $J = \{j\}$ is a cofinite strongly directed set. We shall proceed inductively.

First, suppose that j is an initial object of J. Define a fibrant simplicial sets Z_j and a trivial cofibration $X_j \longrightarrow Z_j$ by functorially factoring the maps $X_j \longrightarrow *$ as in [Q -1, §II.3].

Next, assume inductively that for a given j in J, and all $k < j$, that simplicial sets Z_k, bonding maps $Z_k \longrightarrow Z_\ell$ for $\ell < k$, and trivial cofibrations $X_k \rightarrow Z_k$ have been defined so that:

a) for $\ell < k$ the diagrams

$$\begin{array}{ccc} X_k & \longrightarrow & Y_k \\ \downarrow & & \downarrow \\ X_\ell & \longrightarrow & Y_\ell \end{array}$$

commute;

b) for $m < \ell < k$ the diagrams

$$\begin{array}{ccc} Z_k & \longrightarrow & Z_\ell \\ & \searrow & \downarrow \\ & & Z_m \end{array}$$

commute; and

c) The maps $Z_k \longrightarrow \lim_{\ell < k}\{Z_\ell\}$ induced by (b) are fibrations.

Apply the Quillen factorization to the composite mapping

$$X_j \longrightarrow \lim_{k < j}\{X_k\} \longrightarrow \lim_{k < j}\{Z_k\}$$

to obtain a diagram

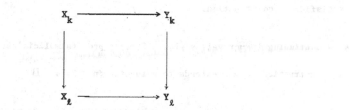

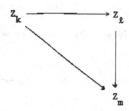

$$X_j \longrightarrow Z_j \longrightarrow \lim_{k<j}\{Z_k\}$$

consisting of a trivial cofibration followed by a fibration. The map $Z_j \longrightarrow \lim_{k<j}\{Z_k\}$ induces bonding maps $Z_j \longrightarrow Z_k$ for $k < j$. It is easy to see that these maps satisfy conditions (a) and (b) above. Condition (c) above is satisfied by construction.

Continuing inductively yields a fibrant pro-(simplicial set) $\{Z_j\} = Ex^\infty\{X_j\}$. By construction, Ex^∞ extends to a functor on SS^J. □

In order to define Ex^∞ on morphisms in pro-SS, we need the following <u>cofinality lemma</u>.

(4.3.3) <u>Lemma</u>. Let $T: J \longrightarrow K$ be a cofinal functor on cofinite strongly directed sets. Then for any $\{X_k\}$, T induces an isomorphism $Ex^\infty\{X_k\} \longrightarrow Ex^\infty\{X_{T(j)}\}$ in Ho(pro-SS), in other words, the diagram

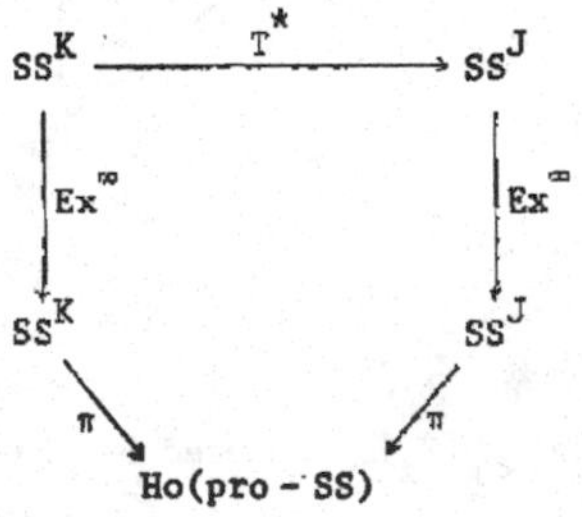

commutes up to natural equivalence of functors.

<u>Proof</u>. Let $\{Y_j\}$ denote $Ex^\infty\{X_{T(j)}\}$ and $\{Z_k\}$ denote $Ex^\infty\{X_k\}$. We may assume that J and K have initial elements j_0 and k_0 with $T(j_0) = k_0$, and

also that $X_{k_0} = *$. This yields a natural map $Z_{T(j_0)} \longrightarrow Y_{k_0}$. Now fix j' in J, and assume that the natural maps $Z_{T(j)} \longrightarrow Y_j$ have been defined for $j < j'$ such that the diagrams

(4.3.4)

$$\begin{array}{ccc} Z_{T(j)} & \longrightarrow & Y_j \\ \downarrow & & \downarrow \\ Z_{T(j'')} & \longrightarrow & Y_{j''} \end{array} \qquad \text{and} \qquad \begin{array}{ccc} X_{T(j)} & \longrightarrow & Z_{T(j)} \\ & \searrow & \downarrow \\ & & Y_j \end{array}$$

commute. Consider the induced commutative diagram

(4.3.5)

$$\begin{array}{ccccc} X_{T(j')} & \longrightarrow & \lim_{k < T(j')}\{X_k\} & \longrightarrow & \lim_{k < T(j')}\{Z_k\} \\ \Big\| & & \downarrow & & \downarrow \\ & & \lim_{T(j) < T(j')}\{X_{T(j)}\} & & \lim_{T(j) < T(j')}\{Z_{T(j)}\} \\ & & \Big\| & & \downarrow \\ X_{T(j')} & \longrightarrow & \lim_{j < j'}\{X_{T(j)}\} & \longrightarrow & \lim_{j < j'}\{Y_j\} \end{array}$$

Since $Z_{T(j')}$ is defined by applying the Quillen factorization to the composite along the top row of diagram (4.3.5), namely the map $X_{T(j')} \longrightarrow \lim_{k < T(j')}\{Z_k\}$, and Y_j, is defined by applying the Quillen factorization to the composite along the bottom row f (4.3.5), $X_{T(j')} \longrightarrow \lim_{j < j'}\{Y_j\}$, there is an induced map

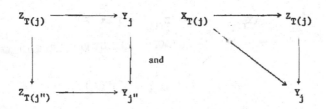

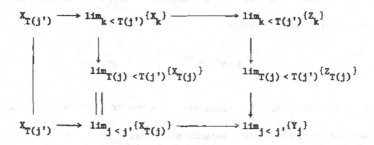

(4.3.6) $$Z_{T(j')} \longrightarrow Y_{j'}.$$

By diagram (4.3.5), the map (4.3.6) satisfies the conditions of diagrams (4.3.4). Continue inductively to define maps $Z_{T(j)} \longrightarrow Y_j$ for all j in J. Diagrams (4.3.4) then yield a commutative diagram in SS^J

(4.3.7)
$$\begin{array}{ccl} \{X_{T(j)}\} & \longrightarrow & \{Z_{T(j)}\} = T^*Ex^\infty\{X_k\} \\ & \searrow & \downarrow \tau \\ & & \{Y_j\} = Ex^\infty T^*\{X_k\} \end{array}$$

Since for each j, $X_{T(j)} \simeq Z_{T(j)}$ and $X_{T(j)} \simeq Y_j$, the map τ in (4.3.7) is a level weak equivalence, hence an isomorphism in Ho(pro - SS).

Since the constructions which led to diagrams (4.3.7) were natural, we have defined the required natural isomorphism

$$\pi \circ Ex^\infty \cong \pi \circ T^* \circ Ex^\infty \longrightarrow \pi \circ Ex^\infty \circ T^*. \quad \square$$

We have thus proven that the construction $X \longrightarrow MX \longrightarrow Ex^\infty MX$, where M denotes the Mardešić construction, extends to a functor from pro - SS to Ho((pro - SS)$_f$) and that there are trivial cofibrations $X \simeq MX \longrightarrow Ex^\infty MX$, which are natural in Ho(pro - SS). By construction, $Ex^\infty M$ factors through Ho(pro - SS), as required.

(4.3.8) Remarks. The only special property of SS which we used is the existence of functorial factorizations in Axiom M2 (§2.3).

§4.4. The derived functors of the inverse limit: background.

In this section we shall briefly summarize vanishing theorems and cofinality theorems for right-derived functors $\lim^s$ of the inverse limit

$$\lim: (AG)^J \longrightarrow AG$$

(where AG is the category of abelian groups). Because lim is left-exact, but not, in general, right exact, vanishing theorems for $\lim^s$ measure the exactness of lim. Cofinality theorems are used to extend $\lim^s$ to pro-AG. There is a close relationship between these results and the homological dimension of modules which we shall not discuss here; see B. L. Osofsky [Osof] for a good survey. We refer the reader to [Mit-2], for example, for the basic theory of derived functors due to H. Cartan and S. Eilenberg [C-E].

Z. Z. Yeh [Yeh] gave the first vanishing theorem in 1959. His results were extended first by J.-E. Roos [Roos] in 1961 and later by C. U. Jensen [Jen] in 1972.

(4.4.1) Definition ([Yeh], [Roos]). An inverse system $\{G_i\}$ indexed by a directed set I is called flasque (or star-epimorphic) if for each ordered (not necessarily directed) set J contained in I, the natural map $\lim_{i \in I}\{G_i\} \longrightarrow \lim_{i \in J}\{G_i\}$ is surjective.

(4.4.2) Proposition ([Yeh], [Roos]). If $\{G_i\}$ is flasque, then $\lim^s\{G_i\} = 0$ for $s \geq 1$.

Roos obtains this result by showing that $\lim^s \{G_i\}$ is the homology of a complex obtained from $\{G_i\}$. Compare J. Milnor's [Mil -3] use of lim and $\lim^1$ in axiomatizing the cohomology of infinite CW complexes. In fact, Roos only requires that I be ordered, not necessarily directed.

(4.4.3) Definition ([Jen]). An inverse system $\{G_i\}$ indexed by a directed set I is called weakly flasque if for each directed set J contained in I, the natural map $\lim_{i \in I}\{G_i\} \longrightarrow \lim_{i \in J}\{G_i\}$ is surjective.

(4.4.4) Proposition ([Jen]). If $\{G_i\}$ is weakly flasque, then $\lim^s\{G_i\} = 0$ for $s \geq 1$.

Jensen first proves that $\lim^s$ applied to a weakly flasque system is 0 by showing that lim is right-exact on such systems. Vanishing of $\lim^s$ then follows by a suitable iteration.

(4.4.5) Remarks.

a) Clearly flasque implies weakly flasque. Jensen observed that the converse is false.

b) Roos and Jensen actually worked in categories of the form A^I, where A is an abelian category which satisfies suitable exactness axioms of A. Grothendieck [Gro -2].

One may extend the domain of lim and $\lim^s$ from $(AG)^J$ to pro-AG as follows. First, a cofinal functor $T:J \longrightarrow K$ induces an isomorphism

$T^*:\{G_k\} \longrightarrow \{G_{T(j)}\}$ in pro - AG, hence we need the following result of Roos (1961), Jensen (1972) and B. Mitchell [Mit - 2] (1973).

(4.4.6) <u>Theorem</u> [Mit-2]. A cofinal functor $T:J \longrightarrow K$ between filtering categories induces commutative diagrams

$$\begin{array}{ccc} (AG)^K & \xrightarrow{\lim_K{}^s} & AG \\ {\scriptstyle T^*}\big\downarrow & \nearrow{\scriptstyle \lim_J{}^s} & \\ (AG)^J & & \end{array}$$

Secondly, although Artin and Mazur [A -M, §A.3] (see §2.1 above) gave a natural representation of a map $\{G_j | j \in J\} \longrightarrow \{H_k | k \in K\}$ in pro - AG by a level map $\{G_{j(\ell)} \longrightarrow H_{k(\ell)}\}$ in some $(AG)^L$, a given map in pro - AG may have many level representatives. Therefore we need to show that commutative diagrams

$$\begin{array}{ccc} \{G_{j(\ell)}\} & \longrightarrow & \{H_{k(\ell)}\} \\ \big\downarrow \cong & & \big\downarrow \cong \\ \{G_{j(m)}\} & \longrightarrow & \{H_{k(m)}\} \end{array}$$

in pro - AG, where $L = \{\ell\}$, and $M = \{m\}$ are filtering categories, induce commutative diagrams

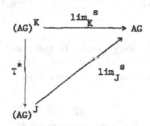

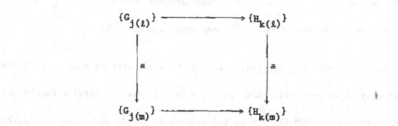

$$\begin{array}{ccc} \lim_L{}^s\{G_{j(\ell)}\} & \longrightarrow & \lim_L{}^s\{H_{k(\ell)}\} \\ \downarrow \cong & & \downarrow \cong \\ \lim_M{}^s\{G_{j(m)}\} & \longrightarrow & \lim_M{}^s\{H_{k(m)}\}. \end{array}$$

J.-L. Verdier [Ver] announced this result in 1965, and hence extended the inverse limit functor and its derived functors to pro-AG. In particular, the diagrams

$$\begin{array}{ccc} (AG)^J & \xrightarrow{\lim_J{}^s} & AG \\ \downarrow & \nearrow_{\lim_{pro}{}^s} & \\ \text{pro-AG} & & \end{array}$$

commute, where $\lim_{pro}{}^s$ is the sth right derived functor of $\lim_{pro}$. We shall include independent proofs of these results, see §4.5.

(4.4.7) <u>lim</u> and <u>HOM</u> <u>functors</u>. Let R be a commutative ring, let R-Mod be the category of R-modules. Let HOM denote the internal mapping functor on R-Mod, that is, HOM (X,Y) is the natural R-module with R-Mod (X,Y) as underlying set. In 1973 B. L. Osofsky [Osof] gave the following representation of the functors $\lim^s : (R\text{-Mod})^J \longrightarrow R\text{-Mod}$. She defined a "<u>tensor</u> <u>product</u>" <u>ring</u> $R \otimes J$ such that the categories $(R\text{-Mod})^J$ and $(R \otimes J)\text{-Mod}$ are isomorphic. This isomorphism induces natural equivalences of functors

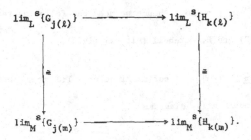

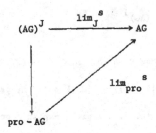

$$\lim: \ (R\text{-Mod})^J \longrightarrow R\text{-Mod}$$

$$(\cong \mathrm{HOM}_{(R\text{-Mod})^J}(R, \lim(-)) : \ (R\text{-Mod})^J \longrightarrow R\text{-Mod})$$

$$\cong \mathrm{HOM}_{(R \otimes J\text{-Mod})}(R \otimes I,-) : \ R \otimes G\text{-Mod} \longrightarrow R\text{-Mod},$$

hence also

$$\lim^s: \ (R\text{-Mod})^J \longrightarrow R\text{-Mod}$$

$$\cong \mathrm{Ext}_{(R \otimes J)\text{-Mod}}{}^s(R \otimes J,-) : \ (R \otimes J\text{-Mod}) \longrightarrow R\text{-Mod}.$$

Later in this chapter we shall use pro-categories to give more natural results See §4.5.

§4.5. Results on derived functors of the inverse limit.

We shall prove the following three theorems in §4.6.

Theorem A. Let J be a cofinite strongly directed set, and let HOM_J, HOM_{pro}, Ext_J and Ext_{pro} be the appropriate HOM and Ext functors. Then

a) $\lim_J = \mathrm{HOM}_J(Z,-): \ AG^J \longrightarrow AG.$

b) $\lim_{pro} = \mathrm{HOM}_{pro}(Z,-): \ \text{pro-}AG \longrightarrow AG.$

c) $\lim_J{}^s = \mathrm{Ext}_J{}^s(Z,-): \ AG^J \longrightarrow AG.$

d) $\lim_{pro}{}^s = \mathrm{Ext}_{pro}{}^s(Z,-): \ \text{pro-}AG \longrightarrow AG.$

Theorem B. Let J be a cofinite strongly directed set. Then the diagrams

$$\begin{array}{ccc} AG^J & \xrightarrow{\pi} & \text{pro}-AG \\ & {\scriptstyle \lim_J{}^s}\searrow \quad \swarrow{\scriptstyle \lim_{pro}{}^s} & \\ & AG & \end{array}$$

commute up to natural equivalence.

Theorem C. Let $\{G_j\}$ be a stable pro-group. Then

$$\lim_J{}^s\{G_j\} = \lim_{pro}{}^s\{G_j\} \cong \lim_{pro}{}^s \lim\{G_j\}) \cong \begin{cases} \lim\{G_j\}, & s = 0, \\ 0, & s \neq 0 . \end{cases}$$

In §4.7 we shall use the relationship between the topological and algebraic structures on pro-SSAG to prove the following.

Theorem D. Let $\{G_j\}$ be an inverse system of free abelian groups and $\{H_k\} \in$ pro-AG. Then

$$\text{Ext}^s(\{G_j\},\{H_k\}) = \text{Ho(pro-SSAG)}(\{G_j\}, \overline{W}^s H_k\})$$

where $\overline{W}$ is the $\overline{W}$-construction of Eilenberg and MacLane (see (4.7.9)).

Theorema A and D imply the following.

Theorem E. $\lim^s\{G_j\} = \text{Ho(pro-SSAG)}(Z, \{\overline{W}^s G_j\}) = \pi_0\{\overline{W}^s G_j\}$ on pro-AG. An analogous formula holds on pro-G for $s = 0$ and 1.

Theorem F. $\lim^s\{G_j\} = \text{Ho(pro-Sp)}(S^0,\{KG_j\})$, where Sp denotes the category of simplicial spectra and KG_j the simplicial Eilenberg-MacLane spectrum with

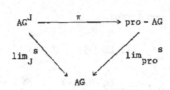

$$\pi_n(KG_j) = \begin{cases} G_j, & n = 0, \\ 0, & n \neq 0 . \end{cases}$$

Bousfield and Kan [B - K, §XI.7] give an analogous formula.

In §4.8 we describe our [E - H - 5] strong-Mittag-Leffler condition for pro-groups, a natural extension of the Mittag-Leffler condition for towers. In [E - H - 5] we proved that a pro-group G is strongly-Mittag-Leffler if and only if G is pro-isomorphic to a flasque pro-group indexed by a cofinite strongly directed set. We give a "topological" proof of the following in §4.8.

Theorem C. Let $\{G_j\}$ be a strongly-Mittag-Leffler pro-group. Then:

a) $\lim^1\{G_j\} = 0$;

b) If $\{G_j\}$ is abelian, then $\lim^s\{G_j\} = 0$ for $s > 0$;

c) If $\lim\{G_j\} = 0$, then $\{G_j\} \cong 0$ in pro-G.

§4.6. Algebraic description of $\lim^s$.

In this section we shall interpret the inverse limit functors $\lim_J : AG^J \longrightarrow AG$ and $\lim_{pro}$: pro-AG $\longrightarrow$ AG as suitable HOM functors. This will show that $\lim_{pro}$ extends $\lim_J$, and also identify the derived functors $\lim^s$ as the derived functors Ext^s of HOM. Compare the results of B. Osofsky described in (4.4.7). Our results can be extended easily to categories of modules over a commutative ring with identity. We shall need the following structure.

(4.6.1) Theorem.

a) AG^J is an abelian category.

b) pro - AG is an abelian category. □

Part (a) is well-known. For part (b), see [A - M, §A.4]. Some of the abelian structure is described below.

(4.6.2) Definition. A sequence $0 \longrightarrow \{A_j\} \longrightarrow \{B_j\} \longrightarrow \{C_j\} \longrightarrow 0$ in AG^J is exact if for each j in J, the sequence $0 \longrightarrow A_j \longrightarrow B_j \longrightarrow C_j \longrightarrow 0$ is exact.

(4.6.3) Proposition. A sequence $0 \longrightarrow \{A_j\} \longrightarrow \{B_k\} \longrightarrow \{C_\ell\} \longrightarrow 0$ in pro - AG is exact if and only if there exists a commutative diagram

(4.6.4)
$$\begin{array}{ccccccccc}
0 & \longrightarrow & \{A_j\} & \longrightarrow & \{B_k\} & \longrightarrow & \{C_\ell\} & \longrightarrow & 0 \\
 & & \downarrow \cong & & \downarrow \cong & & \downarrow \cong & & \\
0 & \longrightarrow & \{A'_m\} & \longrightarrow & \{B_{k(m)}\} & \longrightarrow & \{C'_m\} & \longrightarrow & 0
\end{array}$$

in which the bottom row is a short exact sequence in the appropriate level category AG^M, $M = \{m\}$.

Proof. The "if" part is clear. For the "only if" part, first reindex the given short exact sequence to obtain a diagram

$$\begin{array}{ccccccccc} 0 & \longrightarrow & \{A_j\} & \longrightarrow & \{B_k\} & \longrightarrow & \{C_\ell\} & \longrightarrow & 0 \\ & & \downarrow{\cong} & & \downarrow{\cong} & & \downarrow{\cong} & & \\ 0 & \longrightarrow & \{A_{j(m)}\} & \longrightarrow & \{B_{k(m)}\} & \longrightarrow & \{C_{\ell(m)}\} & \longrightarrow & 0 \end{array} \tag{4.6.5}$$

in which the rows are exact, and the bottom row consists of level maps. Let $M = \{m\}$, and let A'_m be the kernel of the map $B_{k(m)} \longrightarrow C_{\ell(m)}$. Then the map $\{A'_m\} \longrightarrow \{B_{k(m)}\}$ is a kernel of the map $\{B_{k(m)}\} \longrightarrow \{C_{\ell(m)}\}$ in pro - AG, hence we may replace the bottom row of diagram (4.6.5) by the isomorphic short exact sequence

$$0 \longrightarrow \{A'_m\} \longrightarrow \{B_{k(m)}\} \longrightarrow \{C_{\ell(m)}\} \longrightarrow 0. \tag{4.6.6}$$

Similarly, let C'_m be the cokernel of the map $A'_m \longrightarrow B_{k(m)}$. There results a short exact sequence in AG^M,

$$0 \longrightarrow \{A'_m\} \longrightarrow \{B_{k(m)}\} \longrightarrow \{C'_m\} \longrightarrow 0,$$

which is isomorphic to (4.6.6), via reindexing in the middle term, as required. □

(4.6.7) <u>Remarks</u>. Proposition (4.6.3) can easily be extended to finite diagrams of short exact sequences without loops; see §2.1.

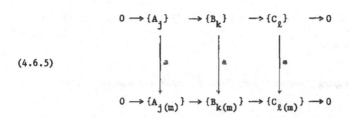

(4.6.8) <u>Definitions</u>. Direct sums in AG^J are defined degreewise. We may define direct sums in pro - AG as follows. Given $\{X_j\}$ and $\{Y_k\}$ indexed by J and K respectively; let $\{X_j\} \oplus \{Y_k\} = \{X_j \oplus Y_k\}$, indexed over the product category $J \times K$. If J and K are directed sets, $(j,k) \leq (j',k')$ if $j \leq j'$ and $k \leq k'$.

The internal mapping functor HOM on AG also extends to AG^J and pro - AG.

(4.6.9) <u>Definitions</u>. Given $\{X_j\}$ and $\{Y_j\}$ in AG^J, define $HOM_J(\{X_j\},\{Y_j\}) = AG^J(\{X_j\},\{Y_j\})$, with group operations induced from $\{Y_j\}$. Similarly, given $\{X_j\}$ and $\{Y_k\}$ in pro - AG, let $HOM_{pro}(\{X_j\},\{Y_k\}) = \text{pro-AG}(\{X_j\},\{Y_k\})$, with group operations induced from $\{Y_k\}$.

Then HOM_J and HOM_{pro} may be extended to functors. Because the inverse limit $\lim_J : AG^J \longrightarrow AG$ (respectively, $\lim_{pro}$: pro - AG $\longrightarrow$ AG) is adjoint to the inclusion $AG \longrightarrow AG^J$ (respectively, AG $\longrightarrow$ pro - AG), and $HOM(Z,X) \cong X$ for X in AG, we have the following.

(4.6.10) <u>Theorem</u>.

a) $\lim_J = HOM_J(Z,-): AG^J \longrightarrow Ag$.

b) $\lim_{pro} = HOM_{pro}(Z,-)$: pro - AG $\longrightarrow$ AG. □

(4.6.11) <u>Corollary</u>. $\lim_J = \lim_{pro} \circ \pi: AG^J \longrightarrow AG$.

(4.6.12) <u>Corollary</u>.

a) $\lim_J{}^s = Ext_J{}^s(Z,-): AG^J \longrightarrow AG$.

b) $\lim_{pro}{}^s = Ext_{pro}{}^s(Z,-)$: pro - AG $\longrightarrow$ AG. □

<u>Proof</u>. In each case, $\lim^s$ and Ext^s are the sth derived functors of isomorphic functors. Now apply D. Buchsbaum's characterization of derived functors [Buch] (see also [Mit-1, p. 193]). The conclusions follows. □

The above theorem and corollaries form Theorem A. We shall now define a natural transformation of <u>connected sequences of functors</u> $\tau:\{\mathrm{Ext}_J{}^s\} \longrightarrow \{\mathrm{Ext}_{pro}{}^s\}$, and show that if J is a cofinite directed set, then τ induces an <u>isomorphism</u> of connected sequences of functors $\tau:\{\mathrm{Ext}_J{}^s(Z,-)\} \longrightarrow \{\mathrm{Ext}_{pro}{}^s(Z,-)\}$.

(4.6.13). <u>Construction of τ</u>. Because the natural quotient functors $\pi:AG^J \longrightarrow \text{pro}-AG$ preserve abelian structures, they map an extension (long exact sequence) in AG^J into an extension in pro-AG, and send a map of extensions in AG^J into a map of extensions in pro-AG. Therefore, τ induces the required natural transformation of connected sequences of functors $\tau:\{\mathrm{Ext}_J{}^s\} \longrightarrow \{\mathrm{Ext}_{pro}{}^s\}$.

(4.6.14) <u>Theorem</u>. Let J be a cofinite strongly directed set. Then the diagrams

$$\begin{array}{ccc} AG^J & \xrightarrow{\quad\pi\quad} & \text{pro}-AG \\ & {\scriptstyle \mathrm{Ext}_J{}^s(Z,-)}\searrow \quad \swarrow{\scriptstyle \mathrm{Ext}_{pro}{}^s(Z,-)} & \\ & AG & \end{array}$$

commute up to natural equivalence.

<u>Proof</u>. For $s = 0$, this follows from (4.6.10) - (4.6.13). The conclusion now follows by Buchsbaum's characterization of derived functors [Buch], or by the

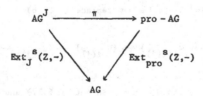

following alternative direct proof for $s > 0$.

We shall first show that $\tau:\mathrm{Ext}_J{}^1(Z, \{G_j\}) \longrightarrow \mathrm{Ext}_{pro}{}^1(Z, \{G_j\})$ is an epimorphism for each cofinite strongly directed set J and $\{G_j\}$ in AG^J. Consider an extension in pro - AG

$$(4.6.15) \qquad 0 \to \{G_j\} \to \{H_k\} \to Z \to 0.$$

By the Mardešić construction, §2.1, we may assume that K as well as J is a cofinite strongly directed set. We shall reindex (4.6.15) several times. First, replace the monomorphism $\{G_j\} \to \{H_k\}$ by an inverse system of monomorphisms $\{G_{j(\ell)}\} \longrightarrow \{H'_\ell\}$, and take the levelwise cokernel $\{C_\ell\}$ to obtain

$$(4.6.16) \qquad 0 \to \{G_{\ell(j)}\} \to \{H'_\ell\} \to \{C_\ell\} \to 0$$

Again, we may assume that $L = \{\ell\}$ is a cofinite strongly directed set. Because J and L are confinite strongly directed sets, we may choose elements $\ell(j)$ for each j in J so that $j \leq j'$ implies $\ell(j) \leq \ell(j')$ and $\{j(\ell(j))\}$ is cofinal in J. We now replace (4.6.16) by

$$(4.6.17) \qquad 0 \to \{G_{j(\ell(j))}\} \to \{H'_{\ell(j)}\} \to \{C_{\ell(j)}\} \to 0.$$

Now use the maps $G_{j(\ell(j))} \to G_j$ to push out (4.6.17) and obtain

$$(4.6.18) \qquad 0 \to \{G_j\} \to \{H''_j\} \to \{C'_j\} \to 0.$$

Finally, the map $Z \cong \{Z_j = Z\} \longrightarrow \{C'_j\}$ is a pro-isomorphism, so pulling back (4.6.18) by this map yields the required extension

(4.6.19) $$0 \longrightarrow \{G_j\} \longrightarrow \{H'''_j\} \longrightarrow \{Z_j\} \longrightarrow 0$$

isomorphic to (4.6.15). Hence, $\tau : \mathrm{Ext}_J^{\,1}(Z, \{G_j\}) \longrightarrow \mathrm{Ext}_{\mathrm{pro}}^{\,1}(Z, \{G_j\})$ is an epimorphism. Similar techniques imply that for each cofinite strongly directed set J and $\{G_j\}$ in AG^J, $\tau : \mathrm{Ext}_J^{\,s}(Z, \{G_j\}) \longrightarrow \mathrm{Ext}_{\mathrm{pro}}^{\,s}(Z, \{G_j\})$ is an isomorphism for all s. The crucial point is that Z is a constant inverse system. □

Because we have already identified $\lim^s$ as Ext^s, Theorem (4.6.14) implies the following.

(4.6.20) <u>Theorem B</u>. Let J be a cofinite strongly directed set. Then the diagrams

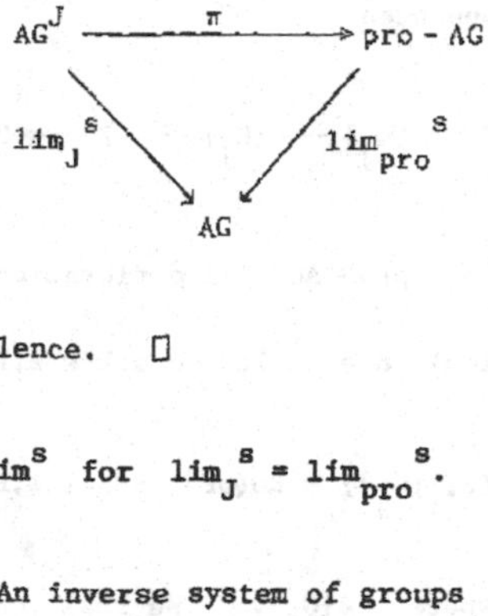

commute up to natural equivalence. □

Hence we shall write $\lim^s$ for $\lim_J^{\,s} = \lim_{\mathrm{pro}}^{\,s}$.

(4.6.21) <u>Definition</u>. An inverse system of groups $\{G_j\}$ is called <u>stable</u> if it is isomorphic in pro-G to a group.

If $\{G_j\}$ is stable, the natural map $\lim \{G_j\} \longrightarrow \{G_j\}$ is an isomorphism in pro-G by functoriality of lim. Theorem B then immediately implies the following vanishing theorem. (If $\{G_j\}$ is not abelian, everything works for

$s = 0$ or 1. See, for example, [B-K] for $\lim^1$ of an inverse system of non-abelian groups. In this case, $\lim^1$ is only a pointed set.)

(4.6.22) Theorem C. Let $\{G_j\}$ be a stable pro-group. Then

$$\lim_J{}^s\{G_j\} = \lim_{pro}{}^s\{G_j\} = \lim_{pro}{}^s(\lim\{G_j\}) \cong \begin{cases} \lim\{G_j\}, & s = 0 \\ 0, & s \neq 0. \end{cases}$$

§4.7. Topological description of $\lim^s$.

In the last section we showed that

$$\lim^1 G_j = Ext^1(Z, \{G_j\})$$

where $J = \{j\}$ is a cofinite strongly directed set. Because $Ext^1(Z, \{G_j\})$ is the set of short exact sequences

$$0 \longrightarrow \{G_j\} \longrightarrow \{H_j\} \longrightarrow \{Z\} \longrightarrow 0,$$

and a short exact sequence in pro-AG is a fibration sequence in pro-SSAG (see (4.7.1) - (4.7.13), below), one is led to ask whether

$$Ext^1(Z, \{G_j\}) = Ho(\text{pro-SSAG})(Z, B\{G_j\})$$

for a suitable classifying space $B\{G_j\}$. We shall carry out the above program in this section. The first step is to relate the abelian structure of pro-AG and the closed model structure of pro-SSAG.

(4.7.1) Definition. Associate to an abelian group G the discrete simplicial abelian group SG with $(SG)_n = G$ for all $n \geq 0$, and all face and degeneracy maps the identity. Associate to a simplicial abelian group H the abelian group $TH = H_0$, the group of 0-simplices of H.

(4.7.2) <u>Proposition</u>. Then S and T extend to functors

$$S : AG \longrightarrow SSAG,$$
$$T : SSAG \longrightarrow AG,$$

with S coadjoint to T, S a full embedding, and $TS = 1_{AG}$.

The proof is easy and omitted.

Now prolong S and T to functors

(4.7.3)
$$S : \text{pro-}AG \longrightarrow \text{pro-}SSAG$$
$$T : \text{pro-}SSAG \longrightarrow \text{pro-}AG$$

(4.7.4) <u>Proposition</u>. S is coadjoint to T, S is a full embedding and $TS = 1_{\text{pro-}AG}$.

<u>Proof</u>. Immediate from Proposition (4.7.2). □

We shall frequently identify pro-AG with its image in pro-SSAG under S. Artin and Mazur [A-M, §A.4] showed that pro-A is an abelian category if A is an abelian category. See §4.6 for the case A = AG. We are in the following situation.

(4.7.5) SSAG is an abelian category. The required structures, namely 0, kernels, cokernels, and direct sums are defined degreewise. Addition in

SSAG (G,H) is defined degreewise. The functors S and T (4.7.2) preserve abelian structures.

(4.7.6) The <u>normalization</u> NG of a simplicial abelian group $G = \{G_n, d^n{}_i, s^n{}_i\}$ is the chain complex $NG = \{N_nG, d_n\}$ with

$$N_nG = \begin{cases} G_0, & n = 0 \\ \bigcap\limits_{i>0} \ker (d^n{}_i : G_n \longrightarrow G_{n-1}), & n = 0, \end{cases}$$

$$d_n = d^n{}_0 \big| N_nG .$$

Then N extends to a functor on SSAG. Moore showed that

$$\pi_* G = H_* \{N_nG, d_n\},$$

the homology of the chain complex NG.

(4.7.7) Call a simplicial abelian group G <u>discrete</u> if G is in the image of $S: AG \longrightarrow SSAG$, that is, if the only non-degenerate simplices of G have dimension 0. If G is a discrete simplicial abelian group, $G_n = G_0$ for all n, and $d^n{}_i = 1_{G_0}$ for all n and i; hence

$$N_nG = \begin{cases} G_0, & n = 0 \\ 0, & n \neq 0 \end{cases}$$

if G is discrete.

Proposition (4.7.6) yields the following relationship between the <u>abelian</u> and <u>closed model structures</u> of SSAG.

(4.7.7) Proposition (Quillen [Q - 1, Proposition II.3.1]). A map $f:G \longrightarrow H$ a fibration in SSAG (hence in SS) if and only if $N_n f$ is surjective for > 0.

(4.7.8) Remarks. We shall need two special cases:

a) Any map of discrete simplicial abelian groups is a fibration;

b) Any (levelwise) surjection of simplicial abelian groups is a fibration.

The fibre of a fibration f in SSAG is just the (levelwise) kernel of f.

(4.7.9) S. Eilenberg and S. MacLane defined a classifying space construction ı SSG (see [May - 1, p. 21] for a description). To a simplicial abelian group their construction associates a fibration sequence in SSAG

$$G \longrightarrow WG \longrightarrow \overline{W}G ;$$

; is a contractible (in SSAG) simplicial abelian group, and $\overline{W}G \in$ SSAG. In ıct WG is always a simplicial group, even if G is not abelian. Their con- :ruction immediately yields the following.

(4.7.10) Proposition. $\overline{W}$ takes a short exact sequence $0 \longrightarrow K \longrightarrow G \longrightarrow H \longrightarrow 0$ ı SSAG into a short exact sequence $0 \longrightarrow \overline{W}K \longrightarrow \overline{W}G \longrightarrow \overline{W}H \longrightarrow 0$ in SSAG. □

(4.7.11) Proposition. A map $f:G \longrightarrow H$ in SSAG is a surjection if and ıly if $\overline{W}f$ is a fibration.

<u>Proof</u>. The "if" part follows from Proposition (4.6.10) and Remarks (4.7.8). For the "only if" part, $(\bar{W}G)_0$ and $(\bar{W}H)_0$ each consist of trivial group 0, so the map $N\bar{W}f : N\bar{W}G \twoheadrightarrow N\bar{W}H$ is a degreewise surjection by Proposition (4.7.7). By [Q-1, Lemma II.3.5], the map $\bar{W}f$ is a surjection. Hence, f is a surjection by the construction of $\bar{W}$. □

We shall now extend the above discussion to pro-SSAG. Call a pro-(simplicial abelian group) <u>discrete</u> if it is pro-isomorphic to one in the image of S: pro-AG $\longrightarrow$ pro-SSAG. Prolong the $\bar{W}$-construction levelwise to pro-SSAG.

(4.7.12) <u>Proposition</u>. All discrete simplicial abelian groups are fibrant.

<u>Proof</u>. If $\{G_j\}$ is discrete, the maps $G_j \longrightarrow \lim_{k<j}\{G_k\}$ are fibrations by Remarks (4.7.8). □

(4.7.13) <u>Proposition</u>. $\bar{W}$ takes a short exact sequence $0 \to K \to H \to G \to 0$ in pro-SSAG into a short exact sequence $0 \to \bar{W}K \to \bar{W}G \to \bar{W}H \to 0$ in pro-SSAG.

<u>Proof</u>. Use Propositions (4.6.3) and (4.7.10). □

The analogue of Proposition (4.7.11) is difficult to state; the ideas will be used in the latter part of this section.

We shall now use pro-SSAG and Ho(pro-SSAG) to classify extensions in pro-AG.

(4.7.14) <u>Proposition</u>. For G and H in pro-AG, S induces a natural isomorphism

$$\sigma:\mathrm{Ext}_{\text{pro}-AG}{}^{s}(G,H) \longrightarrow \mathrm{Ext}_{\text{pro}-SSAG}{}^{s}(G,H).$$

Proof. Because S is full, $\mathrm{HOM}_{\text{pro}-AG}(G,H) = \mathrm{HOM}_{\text{pro}-SSAG}(G,H)$. Also, because S is full and $TS = 1_{\text{pro}-AG}$, the induced natural transformation

$$\sigma:\mathrm{Ext}_{\text{pro}-AG}(G,H) \longrightarrow \mathrm{Ext}_{\text{pro}-SSAG}(G,H)$$

is a monomorphism. To show that σ is an epimorphism, consider an extension

$$E:0 \longrightarrow G \longrightarrow X \longrightarrow \cdots \longrightarrow X' \longrightarrow H \longrightarrow 0$$

in pro-SSAG, where G and H are discrete. Applying T to E yields an extension

$$TE:0 \longrightarrow G \longrightarrow TX \longrightarrow \cdots \longrightarrow TX' \longrightarrow G \longrightarrow 0$$

in pro-AG. Because there is a (natural) map $STE \longrightarrow E$ which is the identity on G and H, $E \cong STE \in \mathrm{Im}\,\sigma$, as required. □

We shall henceforth simply write Ext for $\mathrm{Ext}_{\text{pro}-AG}$ and $\mathrm{Ext}_{\text{pro}-SSAG}$.

(4.7.15) Theorem D. For a levelwise free abelian pro-(abelian group) $\{G_j\}$, and for $\{H_k\}$ in pro-AG,

$$\mathrm{Ext}^{s}(\{G_j\}, \{H_k\}) = \mathrm{Ho(pro-SSAG)}(\{G_j\}, \{W^{s}H_k\}).$$

Proof. First, consider the case $s = 0$, where $\mathrm{Ext}^{0}(\{G_j\}, \{H_k\}) = \mathrm{HOM}\,(\{G_j\}, \{H_k\})$. Because $\{G_j\}$ is free, hence cofibrant (see §2.3, §3.4, and [Q-1, §II.3]), and $\{H_k\}$ is discrete, hence fibrant (Proposition (4.7.12)),

$$\mathrm{Ho(pro-SSAG)}(\{G_j\}, \{H_k\}) = [\{G_j\}, \{H_k\}]$$

(homotopy classes of maps with respect to the cocylinder $\{H_k\}^{[0,1]} = \{H_k{}^{[0,1]}\}$). Because $\{H_k\}$ is discrete, $\{H_k{}^{[0,1]}\} \cong \{H_k\}$. Hence,

$$\mathrm{Ho(pro-SSAG)}(\{G_j\}, \{H_k\}) = \mathrm{pro-SSAG}\,(\{G_j\}, \{H_k\}),$$

as required.

For $s > 0$, we shall use one of B. Mitchell's characterizations of derived functors [Mit-1, p. 198, case III]. Suppose first that $0 \longrightarrow \{A_i\} \longrightarrow \{B_i\} \longrightarrow \{C_i\} \longrightarrow 0$ is a short exact sequence of level maps in pro-SSAG, that is, a short exact sequence in the appropriate level category SSAG^J. Consider a fixed j in J. Then there are fibre sequences

$$A_j \longrightarrow B_j \longrightarrow C_j, \quad \text{and}$$

$$\overline{W}A_j \longrightarrow \overline{W}B_j \longrightarrow \overline{W}C_j .$$

We may obtain a connecting morphism in Ho(SSAG), $\delta : C_j \longrightarrow \overline{W}A_j$ and thus a co-Puppe (fibration) sequence (each map is the fibre of the next map)

$$\text{(4.7.16)} \qquad A_j \longrightarrow B_j \longrightarrow C_j \longrightarrow \overline{W}A_j \longrightarrow \overline{W}B_j \longrightarrow \overline{W}C_j$$

as follows. The homomorphisms $B_j \longrightarrow 0$ and $WA_j \longrightarrow 0$ induce the fibrations f and g in the diagram

$$\begin{array}{ccccc}
 & & WA_j & & \\
 & & \downarrow & & \\
B_j & \longrightarrow & WA_j \times_{A_j} B_j & \xrightarrow{f} & WA_j \times_{A_j} 0 \cong \overline{W}A_j \\
 & & \downarrow g & & \\
 & & 0 \times_{A_j} B_j \cong C_j & &
\end{array}$$

Further, g is an equivalence. Let δ be the composite in Ho(SSAG)

$$C_j \cong 0 \times_{A_j} B_j \xrightarrow{[g]^{-1}} WA_j \times_{A_j} B_j \xrightarrow{[f]} WA_j \times_{A_j} 0 \cong \overline{W}A_j .$$

Then δ is the required connecting morphism, and it is easy to show that sequence (4.7.16) is fibration (co-Puppe) sequence. Therefore, the sequence

$$(4.7.17) \qquad \{A_j\} \longrightarrow \{B_j\} \longrightarrow \{C_j\} \longrightarrow \{\overline{W}A_j\} \longrightarrow \{\overline{W}B_j\} \longrightarrow \{\overline{W}C_j\}$$

is an inverse system of long fibration sequences, hence a long fibration sequence in Ho(pro-SSAG) by Proposition (3.4.17).

For any G in Ho(pro-SSAG), (4.7.17) induces a long exact sequence

$$(4.7.18) \quad [G,\{A_i\}] \longrightarrow [G,\{B_i\}] \longrightarrow [G,\{C_i\}] \xrightarrow{\delta_*} [G,\{\overline{W}A_i\}] \longrightarrow \cdots$$

of abelian groups, where $[-,-] \equiv$ Ho(pro-SSAG)(-,-). Compare (3.4.16).

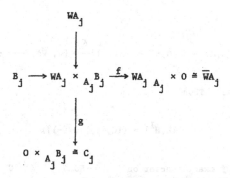

More generally, if $0 \longrightarrow A \longrightarrow B \longrightarrow C \longrightarrow 0$ is a short exact sequence in pro-SSAG, Proposition (4.6.3) and the above techniques yield a long exact sequence

$$(4.7.19) \qquad [G,A] \longrightarrow [G,B] \longrightarrow [G,C] \xrightarrow{\delta_*} [G, \overline{W}A] \longrightarrow \cdots$$

analogous to (4.7.18). Thus

$$(4.7.20) \qquad (H,H^1) = ([G,-],[G,\overline{W}(-)])$$

is a connected pair of exact functor on pro-SSAG. If G is discrete (in pro-AG) and free, and we restrict the functors (4.7.20) to pro-AG, then (H,H^1) is a connected pair of functors with $H = \mathrm{Ho(pro\text{-}SSAG)}(G,-) =$ pro-AG$(G,-)$. Further, $\mathrm{Ho(pro\text{-}SSAG)}(G,\overline{W}(-))$ vanishes on objects of the form $WB = WB_i$ since such objects are levelly contractible (see (4.7.9)). Finally, given any short exact sequence $0 \to A \to B \to C \to 0$ in pro-SSAG, there is a diagram

$$(4.7.21) \qquad \begin{array}{ccccccccc} 0 & \to & A & \to & B & \to & C & \to & 0 \\ & & \| & & \downarrow & & \downarrow & & \\ 0 & \to & A & \to & WB & \to & WB/A & \to & 0 \end{array}$$

in which the bottom row is exact. Thus, sequences of the form $0 \to A \to WB \to WB/A \to 0$ are cofinal in the directed set of all sequences $0 \to A \to B \to C \to 0$. Since $H^1(WB) = 0$, H^1 is the derived functor of H, that is, $H^1(A) = \mathrm{Ext}\,(G,A)$, by Mitchell's criterion.

By iterating the above construction, we obtain the required isomorphisms

$$\mathrm{Ext}^s(\{G_j\},\{H_k\}) = \mathrm{Ho(pro\text{-}SSAG)}(\{G_j\},\{W^s H_k\})$$

for $\{G_j\}$ levelwise free in pro-AG and $\{H_k\}$ in pro-AG. □

The above results hold on pro-G for $s = 0$ and 1. Details are omitted.

We may use the homotopy inverse limit (§4.2) to reformulate the above theorem (for (a) and (b) see Bousfield and Kan [B-K, §XI.7]). Part (c) is our Theorem F (§4.5).

(4.7 ??) **Theorem.**

a) Let $\{G_j\}$ be an inverse system of groups. Then

$$\pi_n(\text{holim}\ \overline{W}G_j) = \begin{cases} \lim^1 \{G_j\} & \text{if } n = 0, \\ \lim \{G_j\} & \text{if } n = 1, \\ 0 & \text{if } n > 1. \end{cases}$$

b) Let $\{G_j\}$ be an inverse system of abelian groups. Then

$$\pi_n(\text{holim}\ \{\overline{W}^s G_j\}) = \begin{cases} \lim^{s-n}\{G_j\} & \text{if } 0 \le n \le s, \\ 0 & \text{if } n > s. \end{cases}$$

c) Let $\{G_j\}$ be an inverse system of abelian groups, and let KG_j be the simplicial spectrum obtained from the simplicial prespectrum $\{\overline{W}^s G_j \mid s \ge 0\}$. Then using stable homotopy groups,

$$\pi_n^s(\text{holim}\ \{KG_j\}) = \begin{cases} \lim^{-n}\{G_j\} & n \le 0, \\ 0 & n > 0. \end{cases}$$

Proof. (a) and (b) follow from the observation

$$\begin{aligned}\pi_n (\operatorname{holim} \overline{W}^s G_j) &= \operatorname{Ho}(\text{pro}-SS)\ (S^n, \{\overline{W}^s G_j\}) \\ &= \operatorname{Ho}(\text{pro}-SS)\ (S^0, \{\overline{W}^{s-n} G_j\}) \\ &= \operatorname{Ho}(\text{pro}-SSAG)\ (Z, \{\overline{W}^{s-n} G_j\}).\end{aligned}$$

together with Theorem E. For (c), use analogous computations with simplicial spectra Sp and simplicial abelian group spectra SpAG [Kan - 1, 2, 3], or use (b) and consider each KG_j as a simplicial prespectrum. □

§4.8. Strongly Mittag-Leffler pro- groups.

We shall give an appropriate generalization to inverse systems indexed by uncountable indexing sets of the following well-known results.

(4.8.1) Let $\{G_n\}$ be a tower of groups such that the bonding maps are all surjections. Then $\lim^s\{G_n\} = 0$ for $s > 0$.

(4.8.2) Suppose further that $\lim \{G_n\} = 0$, then $\{G_n\} \cong 0$ in pro - G.

A pro -group is said to satisfy the Mittag-Leffler (M - L) condition if it is pro- isomorphic to an inverse system of groups whose bounding maps are surjections.

Keesling [Kee - 2] has exhibited a M - L pro - (abelian group) $\{G_i\}$, indexed by an uncountable directed set, such that $\lim \{G_i\} = 0$ but $\{G_i\} \not\cong 0$ in pro - G. Thus (4.8.2) fails in general for M - L pro - groups. Keesling [Kee - 2] also constructed a movable (Definition (4.8.5), below) inverse system of long exact sequences of abelian groups such that the inverse limit sequence is not exact. We shall use this example to prove (Proposition (4.8.6), below) that (4.8.1) also fails, in general, on M - L pro - groups.

In a positive direction, we suggest the following definition as the appropriate generalization of the M - L condition to uncountable inverse systems.

(4.8.3) Definition. A pro - group is said to satisfy the strong-Mittlag-Leffler (S - M - L) condition if it is pro - isomorphic to a pro - group $\{G_j\}$ such that $J = \{j\}$ is a cofinite strongly directed set, and for all j in J, the natural maps $G_j \longrightarrow \lim_{k<j}\{G_k\}$ are surjections.

Clearly, S - M - L implies M - L. Also, for towers, M - L implies S - M - L.

(4.8.4) Proposition [E - H - 5]. A pro - group G is strongly Mittag-Leffler if and only if G is pro - isomorphic to a flasque pro - group indexed by a cofinite strongly directed set.

We sketch a proof for completeness. If $\{G_j\}$ is flasque and $J = \{j\}$ is a cofinite strongly directed set, the natural maps $G_j \longrightarrow \lim_{k<j}\{G_k\}$ are surjections by definition. Conversely, suppose $\{G_j\}$ is indexed by a cofinite strongly directed set and the natural maps $G_j \longrightarrow \lim_{k<j}\{G_k\}$ are surjections. Let J' be an ordered subset of J. It is easy to extend an inverse system $\{g_j \mid j \in J'\}$ to an inverse system indexed by J by induction over $J \setminus J'$. Thus $\{G_j\}$ is flasque. □

(4.8.5) Theorem G. Let $\{G_j\}$ be a strongly-Mittag-Leffler pro - group. Then

a) $\lim^1\{G_j\} = 0$;

b) If $\{G_j\}$ is abelian, then $\lim^s\{G_j\} = 0$ for $s > 0$;

c) If $\lim \{G_j\} = 0$, then $\{G_j\} \cong 0$ in pro-G.

<u>Proof</u>. We may assume that $J = \{j\}$ is a cofinite strongly directed set and that for each j the induced map $p: G_j \longrightarrow \lim_{k<j}\{G_k\}$ is a surjection. Then, the induced maps $\overline{W}p: \overline{W}G_j \longrightarrow \overline{W}(\lim_{k<j}\{G_k\})$ are fibrations by Proposition (4.7.11). Further, since $\overline{W}$ is an adjoint functor (see, for example, [May-1, Theorem 27.1]), $\overline{W}(\lim_{k<j}\{G_k\}) \cong \lim_{k<j}\{\overline{W}G_k\}$. Hence the inverse system $\{\overline{W}G_j\}$ is fibrant (§3.3), that is, the induced maps

$$\overline{W}G_j \longrightarrow \lim_{k<j}\{\overline{W}G_k\}$$

are fibrations (in SSG or SSAG).

By Theorem E (§4.6),

$$\lim^1\{G_j\} = \pi_0\{\overline{W}G_j\} = \pi_0(\text{holim}\ \{\overline{W}G_j\})$$

(see §4.2 for holim). Because $\{\overline{W}G_j\}$ is fibrant,

$$\begin{aligned} \text{holim}\ \{\overline{W}G_j\} &\simeq \lim\ \{\overline{W}G_j\} \quad \text{(see (4.2.11))} \\ &\simeq \overline{W}(\lim\ \{G_j\}), \end{aligned}$$

because $\overline{W}$ is an adjoint. Hence, $\pi_0(\text{holim}\ \{\overline{W}G_j\}) = 0$. Part (a) follows.

The proof of part (b) uses the formula

$$\lim^s\{G_j\} = \pi_0\{\overline{W}^s G_j\}$$

in a similar way. Details are omitted.

For part (c), assume that for some j in J, $G_{j_0} \neq 0$. Choose $g \neq e$ (the identity) in G_j. For $k \leq j_0$ in J, let g_k be the image of g. Otherwise use induction on the number of predecessors of j in J to define an element $\{g_j\}$ in $\lim\{G_j\}$ with $g_{j_0} = g \neq e$. Hence, $\lim\{G_j\} \neq 0$. This contradiction establishes part (c). □

Recall that an object $\{X_j\}$ of pro-C is said to be movable if for each j there exists a $k > j$ such that for all $\ell > k$ there exists a filler in the diagram

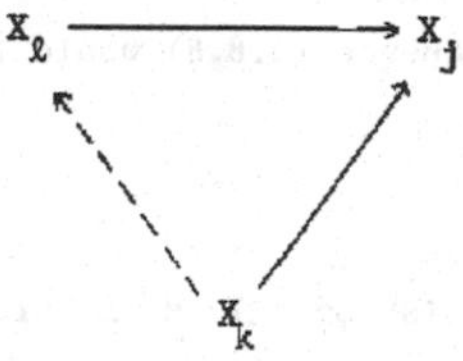

It is easy to check that movable pro-groups satisfy the Mittag-Leffler condition.

(4.8.6) Proposition [E-H-5]. In general, $\lim^1$ need not vanish on Mittag-Leffler pro-groups.

We sketch the proof for completeness. If $\lim^1\{G_j\} = 0$ for all M-L pro-groups $\{G_j\}$, then any short exact sequence

$$0 \longrightarrow \{A_j\} \longrightarrow \{B_k\} \longrightarrow \{C_\ell\} \longrightarrow 0$$

of M-L pro-groups would yield a short exact sequence under the functor lim.

In [Kee-2] Keesling constructs a movable pair, (X_j, A_j) in pro-$\mathrm{ANR}_{\mathrm{pairs}}$, hence a movable system of long exact sequences

$$(4.8.7) \qquad \{\cdots \longrightarrow H_1(X_j) \longrightarrow H_1(X_j,A_j) \xrightarrow{\partial_j} H_0(A_j) \longrightarrow \cdots\},$$

such that the induced sequence

$$(4.8.8) \qquad \lim\{H_1(X_j)\} \longrightarrow \lim\{H_1(X_j,A_j)\} \xrightarrow{\lim\{\partial_j\}} \lim\{H_0(A_j)\}$$

is not exact at the middle term. Since the kernels and images in sequence (4.8.7) are movable, and hence M-L, sequence (4.8.8) would then be exact, contradicting Keesling.

Therefore, $\lim^1$ cannot vanish on all M-L pro-groups.

§4.9. The Bousfield-Kan spectral sequence

In this section we discuss the Bousfield-Kan spectral sequence for the homotopy groups of the homotopy inverse limit of a pro-(simplicial set) [B-K, Chapter XI]. Although their model structure [B-K, p. 314] on SS^J differs from ours, the resulting homotopy categories are isomorphic (see §3.2). Also, if X is a fibrant object in our model structure, then X is fibrant in the Bousfield - Kan structure.

Let $X \in SS^J$. Because our holim preserves weak equivalences, we may replace X by a fibrant object X' with $\text{holim}\, X = \text{holim}\, X'$. Hence, we may assume that X is fibrant. In this case, the Bousfield - Kan homotopy inverse limit, temporarily denoted holim_{B-K} satisfies

$$\begin{aligned} \text{Ho(pro-SS)}(W,X) &\cong \text{Ho}(SS^J)(cW,X) \\ &= \text{Ho(SS)}(W, \text{holim}_{B-K}X) \end{aligned}$$

(cW denotes the constant diagram with $cW_j = W$ for all j in J) [B - K, Proposition XI.8.1]. By uniqueness of adjoints (see, e.g., [Mit - 1]), $\text{holim}\, X \simeq \text{holim}_{B-K}X$. A direct proof also exists, compare Milnor's use of telescopes in [Mil - 3].

Hence the following result of Bousfield and Kan holds for our homotopy inverse limit.

(4.9.1) <u>Theorem</u> [B - K, §XI.7.1, §IX.5.4 - 6]. For a pro - simplicial set X, there is a spectral sequence $\{E_r(X), d_r\}$, with

$$E_2(X) = \{E_2^{p,q}(X) = \lim{}^p_j \{\pi_q(X_j)\}\}, \quad \text{bidegree } d_r = (r, r-1).$$

Under suitable conditions, $\{E_r(X)\}$ converges completely to $\pi_* \text{ holim}\, X$. □

There are two important special cases.

(4.9.2) If X in pro-SS_* is equivalent in pro-$Ho(SS_*)$ to a simplicial set, then

$$E_1^{p,q}(X) = E_\infty^{p,q}(X) = \begin{cases} \pi_q \operatorname{holim} X, & p = 0 \\ 0, & p \neq 0 \end{cases}$$

(4.9.3) If $X = \{X_j\}$ in pro-SS_* is equivalent to a tower $\{X'_n, n = 0,1,\cdots\}$, then $E_2(X) = E_\infty(X)$ and the spectral sequence collapses to the short exact sequences

$$0 \longrightarrow \lim{}^1{}_j\{\pi_{q+1}X_j\} \longrightarrow \pi_q \operatorname{holim} X \longrightarrow \lim{}_j\{\pi_q X_j\} \longrightarrow 0.$$

We shall need the following extensions in order to discuss strong homology (5.6.7) - (5.6.8) and Steenrod homology (§8).

First, we may replace simplicial sets by simplicial spectra in the Bousfield-Kan spectral sequence. Their proof of convergence still applies if suitable care is taken with smash products. In particular, for $\{X_j\} \in$ pro-SS_* and $E \in Sp$, there is a spectral sequence with

$$\begin{aligned} E_2^{p,q} &= \lim{}^p{}_j\{\pi_q{}^S(X_j \wedge E)\} \\ &= \lim{}^p{}_j\{h_q(X_j)\} \end{aligned}$$

which converges under suitable assumptions to

$$\pi_*^S \operatorname{holim} \{X_j \wedge E\}$$

$$\cong {}^S h_*\{X_j\} \quad \text{(see (5.6.8))}$$

$$(\neq h_* (\operatorname{holim} \{X_j\}))$$

Second, suppose $X = \{X_j\}$ where each $X_j = \{X_{j,k}\}$ is itself a pro-(simplicial set) or a pro-(simplicial spectrum). Definition (4.2.10) implies that

$$\operatorname{holim} \{X_{j,k}\} = \operatorname{holim}_j \{\operatorname{holim}_k \{X_{j,k}\}\}.$$

Hence there is a spectral sequence with

$$(4.9.4) \qquad E_2^{p,q} = \lim{}^p{}_j \pi_q (\operatorname{holim}_k \{X_{j,k}\})$$

$$(\cong \lim{}^p{}_j \{\pi_q\{X_{j,k}\}_k\}, \quad \text{see (5.6.1)})$$

§4.10 Homotopy colimits

We shall review the history of homotopy colimits and give a construction dual to our construction of homotopy limits in §4.2.

J. Milnor [Mil-3] introduced the homotopy colimit of a direct tower in his work on axiomatizing additive homology theories on countable CW complexes. Let

$$X = \{X_0 \longrightarrow X_1 \longrightarrow X_2 \longrightarrow \cdots\}$$

be a direct tower. Milnor introduced the (direct) telescope of X

$$\text{Dir tel } X = X_0 \times [0,1] \cup X_1 \times [1,2] \cup X_2 \times [2,3] \cdots,$$

shown below

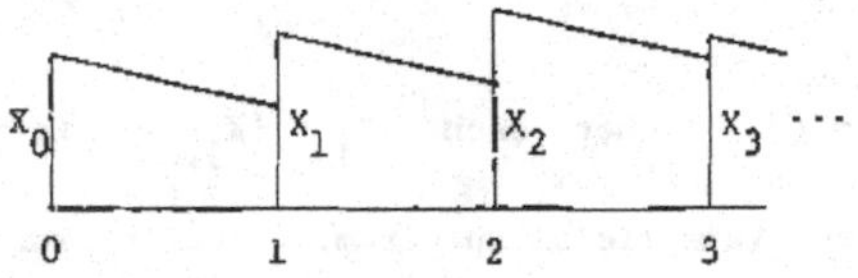

There is a natural map Dir tel X $\longrightarrow$ colim X. Milnor proved that if X is a tower of cofibrations (i.e., X is cofibrant in the category of direct towers or in inj-Top, see §3.8), then this map is a homotopy equivalence.

Bousfield and Kan used a dual construction (replacing the infinite mapping cylinder of a direct tower by the infinite path fibration construction of an (inverse) tower) in their geometric unstable Adams' spectral sequence (see [B-K]). They extended this construction to obtain their holim.

J. Boardman and R. Vogt's work on homotopy everything H-spaces required the homotopy colimit of a diagram; an explicit description is given in [Vogt-1]. This construction is dual to the Bousfield-Kan description.

There is also an evident duality between the Bousfield-Kan spectral sequence and the Segal spectral sequence for the classifying space of a category [Seg].

We obtain a homotopy colimit functor hocolim: Ho(inj-C) $\longrightarrow$ Ho(C) coadjoint to the natural inclusion $Ho(Top) \longrightarrow Ho(inj\text{-}C)$ (C a nice closed model category) by dualizing the construction in §4.2. Explicitly, hocolim X is obtained by replacing X by a cofibrant object X' in inj-C (see §3.8) with $X \simeq X'$ in $Ho(inj\text{-}C)$ and defining

$$\text{hocolim } X \equiv \text{colim } X'.$$

Because the colimit functor colim: $inj\text{-}C \longrightarrow C$ preserves cofibrations and trivial cofibrations (Theorem (3.8.7)),

$$Ho(C)(\text{hocolim } X, Y) \cong Ho(inj\text{-}C)(X,Y)$$

for $X \in inj\text{-}C$ and $Y \in C$. Milnor's theorem on telescopes can be extended to conclude that for X cofibrant in inj-C, there is a natural weak equivalence from the Vogt homotopy colimit to our hocolim.

§5. THE ALGEBRAIC TOPOLOGY OF pro - C.

§5.1. Introduction.

In §3 we showed that if C is a nice model category then so is pro - C. In §4 we developed the theory of homotopy inverse limits for pro - C. In this section our main goal is to describe the algebraic topology of pro - C and to compare Ho(pro - C) with pro - Ho(C).

In §5.2 we prove comparison theorems which relate maps and isomorphisms in Ho(tow - C) to maps and isomorphisms in tow - Ho(C).

§5.3 contains some remarks about completions.

In §5.4 we review the Artin-Mazur theory of pro - Ho(SS_*).

§5.5 is concerned with Whitehead and stability theorems and counter-examples.

In §5.6 we introduce strong homotopy and homology theories and prove a Brown theorem for cohomology theories.

§5.2. Ho(tow - C_*) versus tow - Ho(C_*).

Let C_* be a pointed nice simplicial closed model category which satisfies Condition N of §2.3 and let π:Ho(tow - C_*) $\longrightarrow$ tow - Ho(C_*) denote the natural functor.

(5.2.1) Comparison Theorem. There is a natural short exact sequence of pointed sets:

(5.2.2)
$$0 \longrightarrow \lim^1{}_k \operatorname{colim}_j \{\mathrm{Ho}(C_*)(\Sigma X_j, Y_k)\}$$
$$\longrightarrow \mathrm{Ho}\ (\mathrm{tow}-C_*)(\{X_j\},\{Y_k\})$$
$$\xrightarrow{\pi} \mathrm{tow}-\mathrm{Ho}(C_*)(\{X_j\},\{Y_k\}) \longrightarrow 0.$$

(5.2.3) Remarks. It is easy to see that the natural map $\mathrm{Ho}(\mathrm{tow}-C)(\{X_j\},\{Y_k\}) \longrightarrow \mathrm{tow}-\mathrm{Ho}(C)(\{X_j\},\{Y_k\})$ is always surjective; we need base points only to make a more precise statement.

(5.2.4) Remarks. J. Grossman [Gros -2] obtained the above sequence in his coarser homotopy theory of $\mathrm{tow}-SS_*$.

Proof of Theorem (5.2.1). By Axiom M2 for C_*^{N}, or an easy inductive argument, we can replace any tower by an equivalent (in $\mathrm{Ho}(\mathrm{tow}-C_*)$) fibrant tower, i.e., tower of fibrations of fibrant objects. We may therefore assume that $\{Y_k\}$ is fibrant. Define $Y_{-1} = *$; then the map $Y_0 \longrightarrow Y_{-1}$ is a fibration.

Consider the following function spaces (all of which take values in SS_*; see §2.4):

$$\mathrm{HOM}\ (X_j, Y_k), \quad j \geq 0, \quad k \geq -1;$$

(5.2.5)
$$\mathrm{HOM}\ (\{X_j\}, Y_k) \equiv \operatorname{colim}_j \{\mathrm{HOM}\ (X_j, Y_k)\}, \quad k \geq 0;$$
$$\mathrm{HOM}\ (\{X_j\},\{Y_k\}) \equiv \lim_k \{\mathrm{HOM}\ (\{X_j\}, Y_k)\}.$$

The fibrations $Y_k \longrightarrow Y_{k-1}$ induce fibrations $HOM(X_j,Y_k) \longrightarrow HOM(X_j,Y_{k-1})$ by the simplicial structure on C_*, and also fibrations $HOM(\{X_j\},Y_k) \longrightarrow HOM(\{X_j\},Y_{k-1})$ by the simplicial structure on $\text{pro}-C_*$ or a direct argument (the colimit of a sequence of Kan fibrations is a Kan fibration by the "small-object argument" [Q -1, §II.3]).

Because (5.2.5) expresses $HOM(\{X_j\},\{Y_k\})$ as the limit of a tower of fibrations, the Bousfield-Kan spectral sequence for a tower ((4.9.3)) yields an exact sequence of pointed sets

$$(5.2.6) \qquad 0 \longrightarrow \lim{}^1_k\{\pi_1(HOM(\{X_j\},Y_k))\} \longrightarrow \pi_0(HOM(\{X_j\},\{Y_k\}))$$

$$\longrightarrow \lim_k\{\pi_0(HOM(\{X_j\},Y_k))\} \longrightarrow 0.$$

By the above simplicial structures,

$$\begin{aligned}
\pi_1(HOM(\{X_j\},Y_k)) &\equiv [S^1, HOM(\{X_j\},Y_k)] \\
&\cong \pi_0(HOM(S^1, HOM(\{X_j\},Y_k))) \\
&\cong \pi_0(HOM(\{X_j\} \wedge S^1,Y_k)) \\
&\equiv \pi_0(HOM(\{X_j \wedge S^1\},Y_k)) \\
&\cong \pi_0(HOM(\{\Sigma X_j\},Y_k)) \\
&\cong Ho(\text{pro}-C_*)(\{\Sigma X_j\},Y_k) \\
&\cong \operatorname{colim}_j\ [\Sigma X_j,Y_k]
\end{aligned}$$

([-,-] denotes $Ho(C_*)(-,-)$; the last isomorphism follows by the homotopy extension property). Also,

$$\pi_0(HOM\ (\{X_j\},\{Y_k\}) = Ho(tow-C_*)(\{X_j\},\{Y_k\}), \quad \text{and}$$

$$\pi_0(HOM\ (\{X_j\},Y_k)) = \operatorname{colim}_j\ \{[X_j,Y_k]\},$$

as above. Because the above isomorphisms are natural (§2.4), and $tow-Ho(C_*)(\{X_j\},\{Y_k\}) \cong \lim_k\{\operatorname{colim}_j\{[X_j,Y_k]\}\}$, the conclusion follows from (5.2.6). □

(5.2.7) Remarks. In the above proof, one can replace the Bousfield - Kan spectral sequence by a direct argument dual to Milnor's "$\lim^1$ argument" [Mil -3, Lemma 1]; see, for example [B - K, p. 254].

Theorem (5.2.1) suggests the following.

(5.2.8) Question. Let f be a map in Ho(tow -C) whose image in tow - Ho(C), πf, is invertible. Is f invertible in Ho(tow - C) ?

This question appears quite difficult, and its analogue in proper homotopy theory (see §6.2) has attracted recent interest (Chapmann and Siebenmann [C - S]). At present we can only offer a partial answer (Theorem (5.2.9)). The analogue of (5.2.9) in proper homotopy theory [E -H -3], (see §6.4) answers another question in [C - S].

(5.2.9) <u>Theorem</u>. Let $f:\{X_j\} \longrightarrow \{Y_k\}$ be a map in Ho(tow - C) whose image πf in tow - Ho(C) is invertible. Then there is an isomorphism $g:\{X_j\} \longrightarrow \{Y_k\}$ in Ho(tow - C) with $\pi f = \pi g$ in tow - Ho(C).

<u>Proof</u>. As in the proof of Theorem (5.2.1), we may assume that both $\{X_j\}$ and $\{Y_k\}$ are fibrant. By reindexing if necessary, we may then realize both πf and its inverse in tow - Ho(C) in the following homotopy-commutative diagram over C.

(5.2.10)

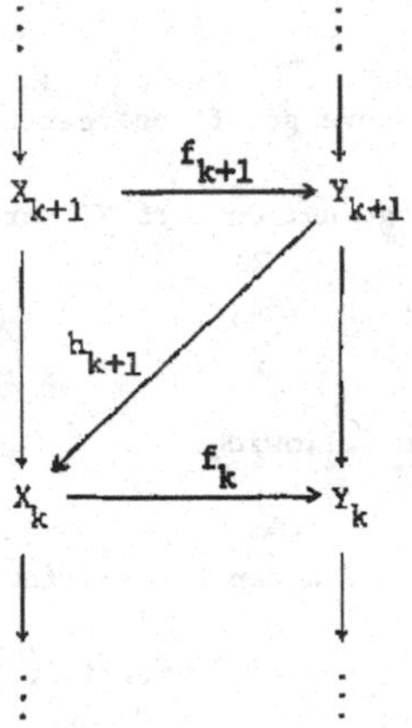

Let Z be the tower

$$\left\{X_0 \xleftarrow{h_1} Y_1 \xleftarrow{f_1} \cdots \xleftarrow{f_k} X_k \xleftarrow{h_{k+1}} Y_{k+1} \xleftarrow{f_{k+1}} X_{k+1} \xleftarrow{h_{k+1}} \cdots\right\}.$$

Form the homotopy commutative diagram

(5.2.11)

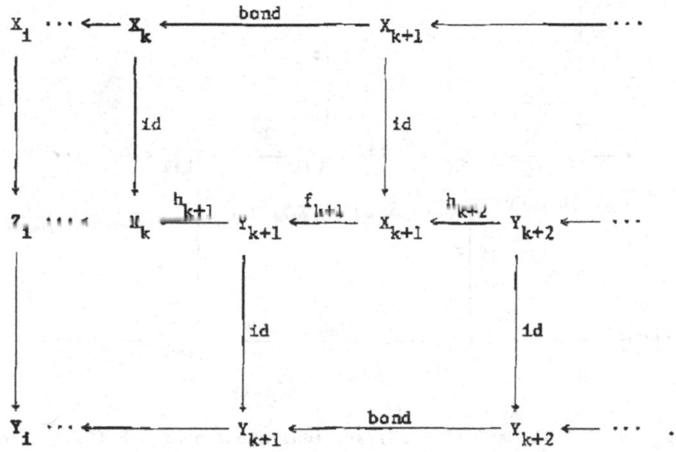

Diagram (5.2.11) factors πf in tow-Ho(C). By inductively deforming the maps $id: X_k \longrightarrow X_k$ and $id: Y_k \longrightarrow Y_k$ above, we may obtain a strictly commutative diagram

(5.2.12)

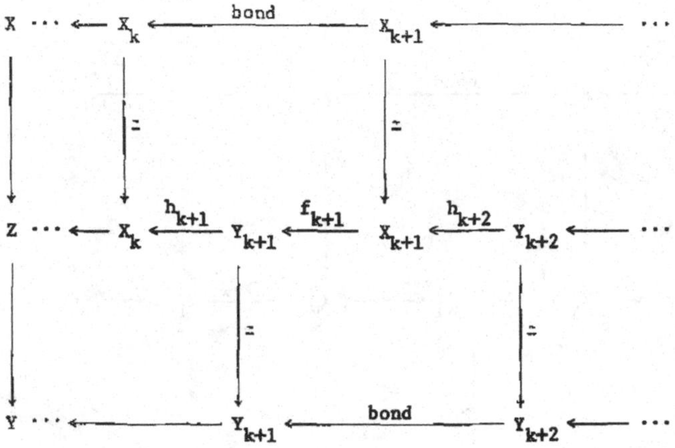

such that diagrams (5.2.11) and (5.2.12) are equivalent over Ho(C). The composite map $g: X \longrightarrow Y \longrightarrow Z$ in diagram (5.2.12) is invertible in Ho(tow - C) because it is the composite of two levelwise weak equivalences, and satisfies $\pi g = \pi f$, as required. □

(5.2.13) <u>Corollary</u>. If either X or Y is stable in tow - Ho(C) (i.e., isomorphic in tow - Ho(C) to an object of Ho(C)), then f above is an isomorphism in Ho(tow - C).

<u>Proof</u>. A first application of the theorem shows that X and Y are stable in Ho(tow -C), say isomorphic to objects X' and Y' of C, respectively. Consider the composite map

(5.2.14) $$X' \xrightarrow{\cong} X \xrightarrow{f} Y \xrightarrow{\cong} Y'.$$

For any Z in C,

$$\begin{aligned} Ho(C)(Z,X') &\cong tow\text{-}Ho(C)(Z,X') \\ &\cong tow\text{-}Ho(C)(Z,Y') \\ &\cong Ho(C)(Z,Y'). \end{aligned}$$

Thus the composite (5.2.14) is an isomorphism in $Ho(C)$, hence in $Ho(tow\text{-}C)$. □

(5.2.15) Remarks. If $Ho(C_*)$ is abelian, the Comparison Theorem implies that f is an isomorphism in $Ho(tow\text{-}C_*)$.

We now make the following observation.

(5.2.16) Rigidification. Each object in $tow\text{-}Ho(C)$ is equivalent to the image of an object in $Ho(tow\text{-}C)$. To see this, given $\{X_j\}$ in $tow\text{-}Ho(C)$, replace $\{X_j\}$ by a tower of fibrant objects $\{X'_j\}$, and choose representatives for the bonding maps of the latter tower.

(5.2.17) Corollary. The isomorphism classification is the same in $tow\text{-}Ho(C)$ and $Ho(tow\text{-}C)$.

Proof. Use (5.2.9) and (5.2.16). □

The above results give a usable relationship between the weak and strong homotopy theories of towers. See §6, especially §6.2, where towers are used in the proper homotopy theory of σ-compact spaces, and §6.5 for a similar application to the shape theory of compact metric spaces.

Off towers, little is known. The Comparison Theorem (5.2.1) is replaced by the following extension of the Bousfield - Kan spectral sequence (see §4.9).

(5.2.16) <u>Theorem</u>. For $\{X_j\}$ and $\{Y_k\}$ in pro $-C_*$, there is a spectral sequence with

$$E^2_{p,q} = \lim{}^p{}_k \{\mathrm{colim}_j \{[\Sigma^q X_j, Y_k]\}\},$$

which is closely related to $\mathrm{Ho}(\mathrm{pro}-C_*)(\{X_j\},\{Y_k\})$.

<u>Proof</u>. We may assume that $\{Y_k\}$ is fibrant. Because $\{j\}$ is filtering,

$$\mathrm{colim}_j \{[\Sigma^q X_j, Y_k]\} = \mathrm{Ho}(\mathrm{pro}-C_*)(\{\Sigma^q X_j\}, Y_k)$$

(each map on the left may be represented by a map $\Sigma^q X_j \longrightarrow Y_k$ for some j). By the simplicial closed model structure on $\mathrm{pro}-C_*$,

$$\mathrm{Ho}(\mathrm{pro}-C_*)(\{\Sigma^q X_j\}, Y_k) \cong \pi_q(\mathrm{HOM}\ (\{X_j\}, Y_k)),$$

and

$$\mathrm{Ho}(\mathrm{pro}-C_*)(\{X_j\},\{Y_k\}) \cong \pi_0(\mathrm{HOM}\ (\{X_j\},\{Y_k\})),$$

where HOM is the "function space" of §3.5. Finally,

$$\mathrm{HOM}\ (\{X_j\},\{Y_k\}) = \lim_k \{\mathrm{HOM}\ (\{X_j\}, Y_k)\}$$

(essentially by "enriched adjunction," §2.4)

$$\simeq \mathrm{holim}_k\ \mathrm{HOM}\ (\{X_j\}, Y_k)$$

(because holim ~ lim on fibrant objects).

Applying the Bousfield-Kan spectral sequence to HOM $(\{X_j\},Y_k)$ gives the desired result. □

Unfortunately, we cannot conclude that

π:Ho(pro-C_*)($\{X_j\},\{Y_k\}$) $\longrightarrow$ pro-Ho(C_*)($\{X_j\},\{Y_k\}$) is onto.

The rigidification question - which objects of pro-Ho(C) "come from" Ho(pro-C) - is unanswered and appears quite hard. Alex Heller asked a similar question about simplicial objects over Ho(SS) and over SS. Also, the isomorphism classification question (compare (5.2.9) - (5.2.15)) is unanswered off towers.

Thus, our present knowledge suffices for the proper homotopy theory of σ-compact spaces (see §6) and the shape theory of compact metric spaces (see §8), but not for more general spaces.

Similarly, there are good comparison theorems relating the strong and weak homotopy theories of direct towers. Also, similarly, the relation between Ho(inj-C) and inj-Ho(C) remains obscure.

§5.3. Remarks on Completions.

In 1965, Artin and Mazur [A-M, Chapter 3] introduced the profinite completion $\hat{X} \in$ pro-Ho(CW_0) of an object $X \in$ Ho(CW_0) in order to prove comparison theorems in étale homotopy theory. The inverse system $\hat{X}$ is indexed by the category which has as objects based homotopy classes of maps $X \longrightarrow \hat{X}_\alpha$ where each X_α has finite homotopy groups, and has as morphisms homotopy commutative triangles

$$\begin{array}{ccc} & X & \\ \swarrow & & \searrow \\ \hat{X}_\alpha & \longrightarrow & \hat{X}_\beta \end{array} .$$

Then the association $(X \longrightarrow \hat{X}_\alpha) \longrightarrow \hat{X}_\alpha$ yields an inverse system $\hat{X}$. Sullivan's work on the Adams conjecture and the homotopy type of spaces such as G/PL led him to study the functor $\lim_\alpha\{[-,\hat{X}_\alpha]\}$ [Sul - 1], [Sul - 2]. By suitably topologizing the functors $[-,\hat{X}_\alpha]$, Sullivan showed that the functor $\lim_\alpha\{[-,\hat{X}_\alpha]\}$ satisfies the Mayer - Vietoris (exactness) axiom as well as the wedge axiom, and hence $\lim_\alpha\{[-,\hat{X}_\alpha]\}$ is representable by Brown's Theorem [Bro]. Sullivan then concentrated on the complex $\overline{X}$ which represented $\lim_\alpha[-,\hat{X}_\alpha]$,

(5.3.1) $$[-,\overline{X}] \cong \lim\ \{[-,\hat{X}_\alpha]\};$$

$\overline{X}$ being a simpler and more familiar object than $\{X_\alpha\}$. In 1972, Bousfield and Kan [B - K], motivated by their work on the Adams spectral sequence, defined for every commutative ring R and pointed simplicial set X a functorial R - completion, $R_\infty X$. They obtain $R_\infty X$ as the simplicial inverse limit of a tower of fibrations $\{R_s X\}$. In this situation it is <u>no</u> longer true that the functors $[-,R_\infty X]$ and $\lim_s\ \{[-,R_s X]\}$ are naturally equivalent. Instead, one has a short exact sequence of pointed sets [B - K] (see §4.9).

(5.3.2) $$0 \longrightarrow \lim{}^1{}_s\{[\Sigma W,R_s X]\} \longrightarrow [W,R_\infty X] \longrightarrow \lim_s\{[W,R_s X]\} \longrightarrow 0.$$

We shall briefly compare (5.3.1) and (5.3.2), modulo rigidification problems (see the end of §5.2). In (5.3.2), we always have

$[W, R_\infty X] \cong Ho(\text{pro}-SS_*)(W, \{R_s X\})$ because $R_\infty X = \text{holim}_s \{R_s X\}$ (see §4.2). In this sense, $[-, R_\infty X] \neq \lim_s \{[-, R_s X]\}$ because $Ho(\text{pro}-SS_*)(-, \{R_s X\}) \neq \lim_s \{[-, R_s X]\}$, and this difference is measured by the $\lim^1$ term of (5.3.2). If $R = Z$, and X is a simply-connected (or even nilpotent [B-K, Chapter 3]) finite complex, then [B-K] $\{Z_s X\}$ is cofinal in $\{X_\alpha\}$. Also, for W finite, the groups $\{[\Sigma W, Z_s X]\}$ are finite, so $\lim^1{}_s \{[\Sigma W, Z_s X]\}$ vanishes. This suggests the following.

(5.3.3) <u>Proposition</u>. Let $\{X_n\}$ be a tower of pointed, connected (SS or CW) complexes. If $\lim_n \{[-, X_n]\}$ is representable and $\lim^1{}_n \{[\Sigma -, X_n]\}$ vanishes, then

$$[-, \text{holim}\,\{X_n\}] \cong \lim_n \{[-, X_n]\}$$

on all pointed complexes.

<u>Proof</u>. Let Q represent $\lim_n \{[-, X_n]\}$. Consider the diagram

(5.3.4)
$$\begin{array}{ccc} [-, Q] & \xrightarrow{\cong} & \lim_n \{[-, X_n]\} \\ & \nwarrow\text{(dashed)} \quad \nearrow & \\ & [-, \text{holim}\,\{X_n\}] & \end{array}$$

Evaluation of the top row on $\text{holim}\,\{X_n\}$ yields the filler which makes (5.3.4) commute. Vanishing of $\lim^1{}_n \{[\Sigma -, X_n]\}$ implies that the <u>group</u> homomorphisms $\pi_i(\text{holim}\,\{X_n\}) \longrightarrow \lim_n \{\pi_i(X_n)\}$, $i \geq 1$, are isomorphisms (see §5.2; also

(5.3.2)). Then diagram (5.3.4) yields isomorphisms $\pi_i(\text{holim}\,\{X_n\}) \approx \pi_i(Q)$, $i \geq 1$. Because vanishing of $\lim^1$ implies holim $\{X_n\}$ is connected, holim $\{X_n\}$ and Q have the same weak (singular) homotopy type by the Whitehead Theorem. The conclusion follows. □

§5.4. Some basic functors.

Artin and Mazur [A-M, §§1-4] introduced the homology and homotopy pro-groups of an object in pro-Ho(SS_0), as well as Postnikov decompositions, the Hurewicz Theorem, and a type of Whitehead Theorem. In this section we shall review the above results, except for the Whitehead Theorem. The Whitehead Theorem will be discussed in §5.5.

Recall that any covariant functor $T:C \longrightarrow D$ may be prolonged to a functor pro-T:pro-$C \longrightarrow$ pro-D. We may therefore define the pro-homotopy and pro-homology functors on pro-Ho(SS_*) by the formulas

(5.4.1)
$$\text{pro-}\pi_i(\{X_j\}) \equiv \{\pi_i(X_j)\},$$
$$\text{pro-}\tilde{H}_i(\{X_j\};A) \equiv \{\tilde{H}_i(X_j;A)\}$$

where A is an abelian group. A similar formula holds for any generalized homology theory. These functors induce pro-homotopy and pro-homology functors on pro-SS_* which satisfy the usual properties with respect to the closed model structure. (Note that fibre sequences and related constructions are not functorial on Ho(SS_*), hence it is difficult to describe homotopy and homology theories on pro-Ho(SS_*).) Artin and Mazur even define homology with twisted coefficients; we shall not need these formulas in our work.

Because cohomology is contravariant, the analogue of formula (5.4.1) for cohomology takes values in the category inj - AG of direct systems of abelian groups. Because colim: inj -AG $\longrightarrow$ AG is exact, Artin and Mazur define the cohomology groups by

(5.4.2) $$\tilde{H}^i(\{X_j\};A) \equiv \mathrm{colim}_J\ \{\tilde{H}^i(X_j;A)\}.$$

The category K_0 of pointed, connected Kan complexes (= fibrant simplicial sets) admits functorial Postnikov-type resolutions (see, e.g., [May -1], or [A -M; §1]). We shall describe these resolutions and the induced Postnikov-type resolutions on pro-Ho(SS_0). Let Δ^n_p denote the p-skeleton of the standard simplicial n-simplex Δ^n. By analogy with the formula $X_n = SS(\Delta^n, X)$ for the set of n-simplices of a simplicial set X, for X in K_0 let $\mathrm{cosk}_p X$ be the simplicial set whose n-simplices are given by

$$(\mathrm{cosk}_p X)_n = SS(\Delta^n_p, X),$$

together with face and degeneracy maps induced by the coface and codegeneracy maps $d^i:\Delta^{n-1}_p \longrightarrow \Delta^n_p$ and $s^i:\Delta^{n+1}_p \longrightarrow \Delta^n_p$ for $0 \leq p \leq n$. Roughly, $\mathrm{cosk}^p_p X$ is obtained from X by adjoining additional n-cells for $n > p$ corresponding to maps from Δ^n_p to X. The inclusions $\Delta^n_p \longrightarrow \Delta^n_{p+1}$ and $\Delta^n_p \longrightarrow \Delta^n$ induce compatible maps $\mathrm{cosk}_p X \longrightarrow \mathrm{cosk}_{p-1} X$ and $X \longrightarrow \mathrm{cosk}_p X$. <u>Caution</u>: these maps are not fibrations. We may define the coskeleta of an arbitrary simplicial set X by the formulas $\operatorname{cosk}_p \mathrm{Ex}^\infty X$ (recall that $\mathrm{Ex}^\infty X$ is a Kan complex naturally weakly equivalent to X). For X in K_0, let $X^{(p)}$ and $X(p)$ be the homotopy - theoretic fibres of the maps below:

$$X^{(p)} \longrightarrow X \longrightarrow \operatorname{cosk}_p X,$$

(5.4.3)

$$X(p) \longrightarrow \operatorname{cosk}_{p+1} X \longrightarrow \operatorname{cosk}_p X.$$

Because SS_0 admits canonical factorizations of maps as trivial cofibrations followed by fibrations ([Q - 1, §II.3], see §4.3), the sequences (5.4.3) are functorial in X. Further, $\operatorname{cosk}_p X$ is $(p-1)$ - connected, and $X(p)$ is an Eilenberg - MacLane space of type $K(\pi_p(X), p)$. We therefore regard the sequences (5.4.3) as the <u>canonical Postnikov resolution</u> of X.

In fact, the above constructions are functorial on $\mathrm{Ho}(SS_0)$; the pth co-skeleton of X is characterized by the properties:

(i) $\pi_i(\operatorname{cosk}_p X) = 0$ for $i \geq p$;

(ii) The canonical map $X \longrightarrow \operatorname{cosk}_p X$ is universal with respect to maps into objects Y with $\pi_i(Y) = 0$ for $i \geq p$.

Similarly, the fibre $X^{(p)}$ is $(p-1)$ connected, the composition $X^{(p)} \longrightarrow X \longrightarrow \operatorname{cosk}_p X$ is trivial, and the map $X^{(p)} \longrightarrow X$ is universal for these properties.

Following Artin and Mazur, we define the Postnikov system of an inverse system $X = \{X_j\}$ in either pro-SS_0 or pro-Ho(SS_0) to be the inverse system

(5.4.4) $$X^{\natural} = \{\operatorname{cosk}_p X_j\}$$

indexed by $\{(p,j)\}$. Clearly $\natural$ extends to functors from pro-SS_0 to pro-SS_0 and pro-Ho(SS_0) to pro-Ho(SS_0). If $\{X_n\}$ is a tower, then

(5.4.5) $$X^{\natural} \cong \{\operatorname{cosk}_n X_n\},$$

so we may restrict (5.4.4) to a functor from towers to towers. A map $f: X \longrightarrow Y$ in pro-Ho(SS_0) (respectively, Ho(pro-SS_0)) is called a $\natural$-isomorphism if it induces an isomorphism on Postnikov systems $f^{\natural}: X^{\natural} \longrightarrow Y^{\natural}$.

By using the above machinery and a spectral sequence argument, Artin and Mazur proved the following.

(5.4.6) Hurewicz Theorem for pro-Ho(SS_0). Let pro-$\pi_i(X) = 0$ for $i < n$, where n is an integer > 1. Then the canonical map

$$\text{pro-}\pi_n(X) \longrightarrow \text{pro-}\tilde{H}_n(X)$$

is an isomorphism of pro-groups.

§5.5. Whitehead and Stability Theorems.

In this section let C be any of SS, SSG, SSAG, Sp (simplicial spectra). Then the Whitehead Theorem (5.5.1) holds in the category C_0 of pointed, connected objects in C.

(5.5.1) <u>Whitehead Theorem in $Ho(C_0)$</u>. A map $f:X \longrightarrow Y$ in $Ho(C_0)$ which induces isomorphisms $f_*:\pi_i(X) \longrightarrow \pi_i(Y)$ for $i \geq 1$ is an isomorphism in $Ho(C_0)$.

A natural question is whether (5.5.1) can be extended to $\text{pro}-Ho(C_0)$ and $Ho(\text{pro}-C_0)$ if the homotopy groups $\pi_i(X)$ $(X \in C_0)$ are replaced by the homotopy pro-groups $\{\pi_i(X_j)\}$ $(\{X_j\} \in \text{pro}-C_0)$. The <u>stability problem</u> (i.e., when is an object of $\text{pro}-Ho(C_0)$ or $Ho(\text{pro}-C_0)$ isomorphic to an object of $Ho(C_0)$) will also be studied using homotopy pro-groups. The following example shows that additional hypotheses are needed for a Whitehead Theorem in pro-homotopy.

(5.5.2) <u>Example</u>. Let $S^\infty \equiv \{\vee_{i \geq n} S^i\}_{n>0}$. Then $\text{pro}-\pi_i(S^\infty) \cong 0$ for all $i \geq 1$, but S^∞ is <u>not</u> equivalent to a point in $\text{pro}-Ho(SS_0)$. In fact, because $\pi_i(S^3) \neq 0$ for infinitely many i, (see e.g., [Spa, Corollary 9.7.6]), there is an essential map $S^\infty \longrightarrow S^3$.

(5.5.3) <u>Whitehead Theorem in $\text{pro}-Ho(C_0)$</u>. Suppose that a map $f:X \longrightarrow Y$ in $\text{pro}-Ho(C_0)$ induces isomorphisms $f_*:\text{pro}-\pi_i(X) \longrightarrow \text{pro}-\pi_i(Y)$ for $i \geq 1$. Then f induces an isomorphism $f^\natural:X^\natural \longrightarrow Y^\natural$ of Postnikov systems in $\text{pro}-Ho(C_0)$. Under either of the following additional conditions, f is an isomorphism in $\text{pro}-Ho(C_0)$:

(a) $\sup_{j,k}\{\dim (X_j),\ \dim (Y_k)\} < \infty$;

(b) For each j, $\dim (X_j) < \infty$, for each k, $\dim (Y_k) < \infty$, and f is movable.

<u>Proof</u>. For $C = SS$, the first part is due to Artin and Mazur [A - M, Corollary (4.4)]. Their proof uses a spectral sequence argument and easily extends to the other C.

Similarly, for $C = SS$, the second part is due to the first author and Geoghegan [E - G - 1,2]; see also [A - M; Theorem (12.5)], [Mos - 1], [Mar - 1], [Mor - 1]. We shall sketch the proof. By reindexing as in §2.1, we may assume that f is a level map $\{f_j : X_j \longrightarrow Y_j\}$ indexed by a directed set. For case (a), let $n = \sup_j\{\dim (X_j), \dim (Y_j)\}$. Consider a fixed j. Choose j_1 such that the diagram

$$\begin{array}{ccc} \pi_1(X_{j_1}) & \longrightarrow & \pi_1(Y_{j_1}) \\ \downarrow & \swarrow & \downarrow \\ \pi_1(X_j) & \longrightarrow & \pi_1(Y_j) \end{array}$$

admits a filler. Convert f_j and f_{j_1} into cofibrations f'_j and f'_{j_k}.

Then the map $\pi_1(Y'_{j_1}, X'_{j_1}) \longrightarrow \pi_1(Y'_j, X'_j)$ is 0. Then the composite map

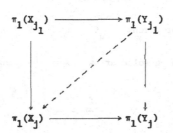

$$Y'^{\,1}_{j_1} \hookrightarrow Y'_{j_1} \longrightarrow Y'_j$$

$(Y'^{\,1}_{j_1}$ denotes the 1-skeleton of $Y'_{j_1})$ may be "deformed" relative to X'_{j_1} into X'_j, similarly for all k with $k \geq j_1$. Now choose $j_2 \geq j_1$, $j_3 \geq j_2, \cdots, j_n \geq j_{n-1}$ so that similar results hold for $\pi_2, \pi_3, \cdots, \pi_n$. We conclude that the map $y'_{j_n} \longrightarrow Y'_j$ can be deformed relative to X'_{j_n} into X'_j (because $\dim(Y_{j_n}) \leq n$). This argument is due to Mardešić [Mar -1] and yields a homotopy inverse to f in case (a) by using the following lemma of K. Morita.

(5.5.4) Lemma. [Mor - 1]. Let $X = \{X_\lambda, p_{\lambda\lambda'}, \Lambda\}$ and $Y = \{Y_\lambda, q_{\lambda\lambda'}, \Lambda\}$ be inverse systems in a category C over the same directed set Λ, and let $\{X_\lambda \xrightarrow{f_\lambda} Y_\lambda\}$ be a level morphism in pro-C. Then f is an isomorphism in pro-C iff for any $\lambda \in \Lambda$ there is some $\mu \in \Lambda$ such that $\lambda \leq \mu$ and there exists $\psi_{\lambda\mu}: Y_\mu \longrightarrow X_\lambda$ for which $\psi_{\lambda\mu} f = p_{\lambda\mu}$ and $f_\lambda \psi_{\lambda\mu} = q_{\lambda\mu}$; i.e., a filler exists in the following solid arrow commutative diagram

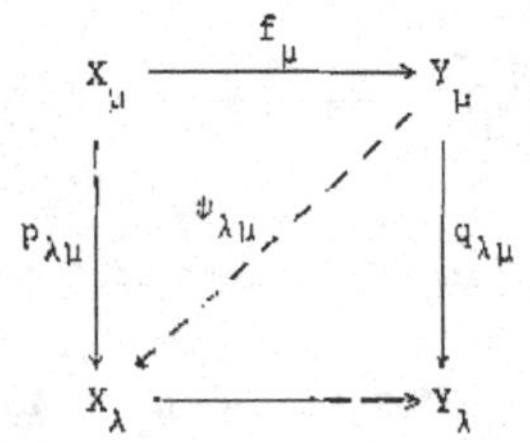

.

Case (a) follows.

In case (b), consider a fixed j. There exists a $k \geq j$ such that for all $\ell \geq k$, homotopy fillers exist in the diagram

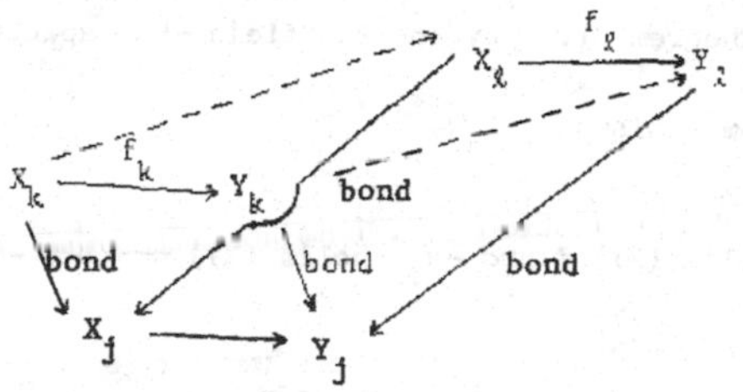

Now use the above argument with $n = \dim(Y_k, X_k)$ to obtain a deformation of the map $Y_k \longrightarrow Y_j$ into X_j relative to X_k. As above, the conclusion follows. □

An object $X \in$ pro-C is said to be <u>stable</u> if it is isomorphic in pro-C to an object of C. The <u>stability problem</u> is the problem of giving criteria on X which imply that X is stable. If $X \in$ pro-Ho(C_0) is stable, then so are its homotopy pro-groups pro-$\pi_i(X)$.

(5.5.5) <u>The Stability Theorem in pro-Ho(C_0)</u> [E-G-1]. Let $X \in$ pro-C_0 and let h:holim$(X) \longrightarrow X$ be the canonical map in pro-Ho(C_0). If pro-$\pi_i(X)$ is stable for all $i \geq 1$, then $h^\natural$ is an isomorphism. h is an isomorphism if X satisfies either of the following conditions:

(a) $\sup_j\{\dim(X_j)\} < \infty$;

(b) X is dominated in pro-Ho(C_0) by an object of Ho(C_0).

We sketch the proof. Because pro-$\pi_i(X)$ is stable, $\lim^s\{\pi_i(X_j)\} = 0$ for $s > 0$ (§4.5, Theorem C). By the Bousfield-Kan spectral sequence (see §4.9), h induces isomorphisms

$$\pi_i(\text{holim}(X)) \cong \text{pro-}\pi_i(\text{holim}(X)) \longrightarrow \text{pro-}\pi_i(X).$$

Therefore $h^\natural$ is an isomorphism by Theorem (5.5.3).

To show that h is an isomorphism in case (a), the first author and Geoghegan applied Wall's finite-dimensionality criterion [Wall] to the homology and cohomology groups of holim (X) (which are isomorphic to the homology pro-groups and cohomology groups of the finite-dimensional system X because $h^\natural$ is an isomorphism by [A-M]). This shows that holim (X) has the homotopy type of a finite-dimensional simplicial set. The conclusion then follows by Theorem (5.5.3).

In case (b), let $u:X \longrightarrow Y$ and $d:Y \longrightarrow X$ be the domination maps with Y in C_0, and $du = 1_X$. One then applies Brown's Theorem [Bro] to split the homotopy idempotent $ud:Y \longrightarrow Y$ ($(ud)^2 = u(du)d = ud$) through Z. One easily checks that

$$X \simeq \{Y \longleftarrow Y \longleftarrow Y \longleftarrow \cdots\} \simeq Z.$$

Then the ordinary Whitehead Theorem implies that the composite map holim $X \longrightarrow Z$ is an isomorphism. The conclusion follows.

(5.5.5.a) <u>Remarks</u>. The argument given in part (b) shows that if X is only assumed to be in $\text{pro}-\text{Ho}(C_0)$, it still follows that X is stable. Dydak has recently shown that the same conclusion holds in part (a).

So far, we have only been able to prove the following strong tower versions of Theorems (5.5.3) and (5.5.5).

(5.5.6) <u>The Whitehead Theorem in $\text{Ho}(\text{tow}-C_0)$</u> Suppose $f:X \longrightarrow Y$ in $\text{tow}-C_0$ induces isomorphisms $f_*:\text{pro}-\pi_i(X) \xrightarrow{\cong} \text{pro}-\pi_i(Y)$ for all $i \geq 1$. Then f induces an isomorphism $f^{\natural}:X^{\natural} \longrightarrow Y^{\natural}$ in $\text{Ho}(\text{tow}-C_0)$. f is itself an isomorphism in $\text{Ho}(\text{tow}-C_0)$ if f satisfies either of the following additional conditions:

a) $\sup\{\dim(X_j), \dim(Y_k)\} < \infty$;

b) f is movable.

We sketch the proof. For $C = SS$ the first part of this theorem is due to Grossman [Gros-2]. For other C, the proof is similar and omitted. The proof of the second part follows from §3.7 and appropriate filtered Whitehead theorems, which are proved in an identical manner to the proper Whitehead Theorem occurring in [Br].

(5.5.7) <u>The Stability Theorem in $\text{Ho}(\text{tow}-C_0)$</u>. Let $X \in \text{tow}-C_0$ and let $h:\text{holim}(X) \longrightarrow X$ be the canonical map in $\text{Ho}(\text{tow}-C_0)$. If $\text{pro}-\pi_i(X)$ is stable for all $i \geq 1$, then $h^{\natural}$ is an isomorphism in $\text{Ho}(\text{tow}-C_0)$. h is an isomorphism in $\text{Ho}(\text{tow}-C_0)$ if X satisfies either of the following conditions:

a) $\sup \{\dim (X_j)\} < \infty$;

b) X is dominated in $\text{pro-Ho}(C_0)$ by an object of $\text{Ho}(C_0)$.

Proof. The first part follows from the first part of (5.5.6). For the second part, Theorem (5.5.5) implies that X is isomorphic in $\text{tow-Ho}(C_0)$ to an object Q of $\text{Ho}(C_0)$. Theorem (5.2.9) then implies that X is isomorphic to Q in $\text{Ho}(\text{tow-}C_0)$. The properties of the functor holim now implies that h is an isomorphism in $\text{Ho}(\text{tow-}C_0)$. □

(5.5.8) Remarks. Porter [Por-2] has given a simple argument which shows that if $X \in \text{Ho}(\text{pro-}SS_0)$ is dominated in $\text{Ho}(\text{pro-}SS_0)$ by an object in $\text{Ho}(SS_0)$, then h is an isomorphism in $\text{Ho}(\text{pro-}SS_0)$. In fact, Porter's argument depends only on the functorial properties of holim, and hence shows that if $X \in \text{Ho}(\text{pro-}C)$ is dominated in $\text{Ho}(\text{pro-}C)$ by an object of $\text{Ho}(C)$, then the canonical map $h:\text{holim}(X) \longrightarrow X$ is an isomorphism in $\text{Ho}(\text{pro-}C)$. If $X \in \text{pro-Ho}(C)$ is dominated by $P \in \text{Ho}(C)$ (i.e., we are given maps $P \underset{u}{\overset{d}{\rightleftarrows}} X$ in $\text{pro-Ho}(C)$ with $du = 1_X$ in $\text{pro-Ho}(C)$), then X is isomorphic in $\text{pro-Ho}(C)$ to $Y = \{P \xleftarrow{ud} P \xleftarrow{ud} \cdots\}$. Y is easily seen to be dominated in $\text{tow-Ho}(C)$ by P. Recent work of Chapman and Ferry [C-F] seems to imply that if $Y \in \text{tow-}C$ is dominated in $\text{tow-Ho}(C)$ by a $P \in \text{Ho}(C)$, then Y is also dominated by P in $\text{Ho}(\text{tow-}C)$, and hence stable by Porter's argument. Thus, X will also be stable in $\text{pro-Ho}(C)$. Thus, an object of $\text{pro-Ho}(C)$ is stable if and only if it is dominated by a stable object.

A map $f:P \longrightarrow P$ is said to be a <u>homotopy idempotent</u> if $f^2 = f$ in Ho(C). f is said to split through Q if there exists maps $P \underset{u}{\overset{d}{\rightleftarrows}} Q$ such that $du = 1_Q$ and $ud = f$ in Ho(C). The above shows that <u>every</u> homotopy idempotent splits. This may be used to show that every homotopy idempotent on an ℓ^2-manifold is homotopic to a strict idempotent and every homotopy idempotent f on a compact Q-manifold M is homotopic to a strict idempotent if and only if a certain Wall obstruction $W(f) \in \tilde{K}_0(\pi_1(M))$ vanishes.[+]

The development of coherent prohomotopy theory should enable one to extend Theorems (5.5.6) and (5.5.7) to Ho(pro-C_0) (see [Por-3]).

Below we present a number of examples which show the precision of the above results.

(5.5.9) <u>Example</u>. Example (5.5.2) showed the need of some extra condition, such as conditions a) or b) of (5.5.3), in order that a ϕ-isomorphism be an isomorphism. In this example we show that it is insufficient to require X and Y to be movable. Such an example was first constructed by Draper and Keesling in [D-K].

Let $S_n = \vee_{i \geq n} S^i$ and $i_n : S_n \longrightarrow S_{n-1}$ be the natural inclusion. Let

---

[+]These observations were obtained in a conversation with T. Chapman.

$X_n = \Pi_{k \leq n} S_k = Y_n$. Let $p_n : X_n \longrightarrow X_{n-1}$, $b_n : Y_n \longrightarrow Y_{n-1}$, and $f_n : X_n \longrightarrow Y_n$ be defined by the following diagram

$$\begin{array}{ccccccc} S_1 \times \cdots \times S_{n-1} \times S_n & \xrightarrow{1 \times \cdots \times 1 \times 0} & S_1 \times \cdots \times S_{n-1} \times S_n \\ \Big\downarrow 1 \quad \cdots \quad \Big\downarrow 1 \;\swarrow i_n & & \Big\downarrow 1 \quad \cdots \quad \Big\downarrow 1 \;\swarrow 0 \\ S_1 \quad \cdots \quad S_{n-1} & \xrightarrow{1 \times \cdots \times 0} & S_1 \times \cdots \times S_{n-1} \end{array} .$$

Then $X = \{X_n, p_n\}$ and $Y = \{Y_n, b_n\}$ are movable and $f = \{f_n\}$ is a $\natural$ - isomorphism (easy), but f is not an isomorphism in pro - Ho(CW_0); in fact, there is an obvious essential map $S^\infty \longrightarrow X$ such that the composition $S^\infty \longrightarrow X \xrightarrow{f} Y$ is inessential.

(5.5.10) <u>Examples</u>. We will show in the examples below that the phenomena exhibited by examples (5.5.2) and (5.5.9) can be realized by inverse systems of <u>finite</u> complexes. These are much deeper examples, though all the depth comes from J. F. Adams. Recall Adams' essential map of Moore spaces Y (these are pointed finite complexes of the form $S^k \cup_q D^{k+1}$) [Adams - 2]

$$A : \Sigma^{2r} Y \longrightarrow Y ;$$

this map is detected by the isomorphism

$$A^* : \tilde{K}(Y) \xrightarrow{\approx} \tilde{K}(\Sigma^{2r} Y) \neq 0,$$

where $\tilde{K}$ denotes reduced K - theory. Hence, the composite maps

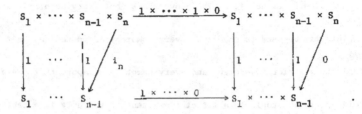

$$A \circ \cdots \circ \Sigma^{2rm-2r} A : \Sigma^{2rm} Y \longrightarrow \Sigma^{2rn} Y$$

are all essential. Thus the inverse system $Z \equiv \{\Sigma^{2rn} Y\}$ bonded by suspensions of A is not equivalent to a point in pro-Ho(CW_0) even though pro-$\pi_i(Z) = 0$ for all $i \geq 1$. Applying the construction of Example (5.5.11) to Z in place of S^∞ yields a map between movable towers of finite complexes which is a $\natural$-isomorphism but not an isomorphism in pro-Ho(CW_0).

The following example provides counter-examples to many conjectures (see [E-H-4]).

Let $\{B_n\}$ be the inverse system with B_0 a point, and $B_n = \Pi_{i=1}^{n} S^{2r}$ for $n \geq 1$, and with bonding maps $B_n \longrightarrow B_{n-1}$ given by projection onto the <u>last</u> $n-1$ S^{2r} factors.

Let $E_n = Y \times B_n$, and let $p_n : E_n = Y \times B_n \longrightarrow B_n$ be the projection; thus F_n, the fibre of p_n, is Y. We shall define "twisted" bonding maps $f_n : E_n \longrightarrow E_{n-1}$ so that the restrictions $f_n|F_n : F_n \longrightarrow F_{n-1}$ are null-homotopic, yet no composite $E_n \longrightarrow E_{n-k}$ factors through a B_{n-1}.

Form the commutative solid-arrow diagrams

$$
\begin{array}{ccccc}
E_n & & E'_n & & E_{n-1} \\
\Vert & & \Vert & & \Vert \\
Y \times (S^{2r})^n & \xrightarrow{\phi_n} & Y \times (S^{2r})^n & \longrightarrow & Y \times (S^{2r})^{n-1} \\
\Big\downarrow {\scriptstyle p_n} & & \Big\downarrow {\scriptstyle p'_n} & & \Big\downarrow {\scriptstyle p_{n-1}} \\
(S^{2r})^n & \longrightarrow & (S^{2r})^n & \longrightarrow & (S^{2r})^{n-1} \\
\Vert & & \Vert & & \Vert \\
B_n & & B_n & & B_{n-1}
\end{array}
$$

(E'_n is the pullback, and p'_n is the projection). Define the filler ϕ_n by requiring that $p'_n\phi_n = p_n$, and letting the composite mapping

$$Y \times (S^{2r})^n \overset{\phi_n}{\dashrightarrow} Y \times (S^{2r})^n \longrightarrow Y$$

be the composite mapping

$$Y \times (S^{2r})^n \xrightarrow{id \times \pi_n} Y \times S^{2r} \longrightarrow Y \wedge S^{2r} \xrightarrow{A} Y ;$$

here π_n is the projection onto the <u>first</u> S^{2r} factor.

Finally, let the bonding map $f_n : E_n \longrightarrow E_{n-1}$ be given by the composite $E_n \xrightarrow{\phi_n} E'_n \longrightarrow E_{n-1}$. This yields the tower $\{E_n\}$ and tower of fibrations $\{F_n \longrightarrow E_n \xrightarrow{p_n} B_n\}$.

<u>Claim 1</u>. The tower F_n is contractible; since the bonding maps $f_n|_{F_n} : F_n \longrightarrow F_{n-1}$ are given by the composites

$$F_n = Y \longrightarrow Y \times * \longrightarrow Y \times S^{2r} \longrightarrow Y \wedge S^{2r} \xrightarrow{A} Y ,$$

$\{F_n\}$ is isomorphic in Pro - Top to $*$.

We may use the basepoints in the $F_n(=Y)$ to define a <u>section</u> $\{s_n : B_n \longrightarrow E_n\}$. Since $\{p_n\}\{s_n\} = id_{\{B_n\}}$, to show that $\{p_n\}$ is not invertible, it suffices to verify the following.

<u>Claim 2.</u> $\{s_n\}\{p_n\} \neq id_{\{E_n\}}$ in Pro - Ho(Top). Assume otherwise, then for arbitrarily large n and suitable m (depending upon n, but with $n - m \geq 0$) the diagram

$$\begin{array}{ccccc}
E_n = Y \times (S^{2r})^n & \xrightarrow{p_n} & (S^{2r})^n & \xrightarrow{s_n} & Y \times (S^{2r})^n \\
\downarrow & & \downarrow & & \downarrow \\
Y \times (S^{2r})^{n-m} & \xrightarrow{p_{n-m}} & (S^{2r})^{n-m} & \xrightarrow{s_{n-m}} & Y \times (S^{2r})^{n-m}
\end{array}$$

with the bond $E_n \longrightarrow E_{n-m}$ and the map $Y \times (S^{2r})^{n-m} \xrightarrow{=} E_{n-m}$

would commute up to homotopy.

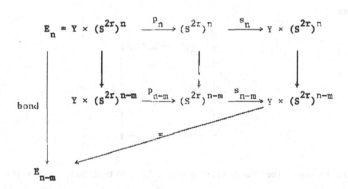

Consider the subdiagram

$$\begin{array}{ccc}
Y \times (S^{2r})^{n} & \xrightarrow{\text{bond}} & Y \times (S^{2r})^{n-m} \\
\Big\downarrow \text{bond} & & \Big\downarrow p_{n-m} \\
 & & (S^{2r})^{n-m} \\
 & & \Big\downarrow s_{n-m} \\
Y \times (S^{2r})^{n-m} & \longrightarrow & Y \times (S^{2r})^{n-m}
\end{array} .$$

Note that all of the above maps are products with $\mathrm{id}_{(S^{2r})^{n-m}}$. Hence, by this fact, and projecting the lower right corner onto Y, we obtain a homotopy commutative diagram

$$\begin{array}{ccc}
Y \times (S^{2r})^{m} & \xrightarrow{g} & Y \\
\Big\downarrow g & & \Big\downarrow \\
 & & * \\
 & & \Big\downarrow \\
Y & \xrightarrow{=} & Y
\end{array}$$

where g is induced from the bonding maps. By construction, g is the composite

$$Y \times (S^{2r})^{m} \longrightarrow Y \wedge (S^{2r})^{m} = \Sigma^{2rm} Y \xrightarrow{A^{m}} Y .$$

Since $\tilde{K}(Y \wedge (S^{2r})^{m})$ is a direct summand in $\tilde{K}(Y \times (S^{2r})^{m})$, $g^{*}: K(Y) \longrightarrow \tilde{K}(Y \times (S^{2r})^{m})$ is non-zero, hence Claim 2 holds. $\square$

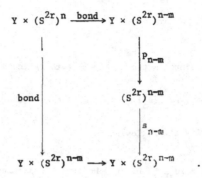

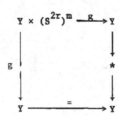

Let $\pi_* \equiv \prod_{i=1}^{\infty} \pi_i \equiv [\bigvee_{i=1}^{\infty} S^i, -]$.

(5.5.10a) Proposition. $\text{pro}-\pi_*\{p_n\}:\{\pi_*(E_n)\} \longrightarrow \{\pi_*(B_n)\}$ is a pro-isomorphism.

Proof. This follows from chasing in the commutative solid arrow diagram

$$\begin{array}{ccccccc}
\pi_*(F_n) & \longrightarrow & \pi_*(E_n) & \overset{s_{n*}}{\underset{pn_*}{\rightleftarrows}} & \pi_*(B_n) & \overset{\partial}{\longrightarrow} & \pi_{*-1}(F_n) \\
\downarrow 0 & & \downarrow & & \downarrow & & \downarrow 0 \\
\pi_*(F_{n-1}) & \longrightarrow & \pi_*(E_{n-1}) & \overset{p_{n-1*}}{\longrightarrow} & \pi_*(B_{n-1}) & \overset{\partial}{\longrightarrow} & \pi_{*-1}(F_{n-1}).
\end{array}$$ □

(5.5.10b) Proposition. An infinite dimensional Whitehead Theorem fails in shape theory.

Proof. Let $\overline{E} = \lim E_n$, $\overline{B} = \lim B_n$; then $p:\overline{E} \longrightarrow \overline{B}$ is a map of compact metric spaces which is an isomorphism on Čech pro-homotopy groups $(\text{pro}-\pi_*)$ but not a shape equivalence. □

(5.5.10c) Remarks. $\overline{p}$ is even a C-E map, see J. L. Taylor [Tay-1].

(5.5.10d) Proposition. An infinite dimensional Whitehead Theorem fails in proper homotopy theory.

Proof. Let Tel E be the infinite mapping cylinder (telescope) Tel $(* \longleftarrow E_0 \longleftarrow E_1 \longleftarrow \cdots)$; define Tel B and Tel p: Tel E $\longrightarrow$ Tel B similarly. Then Tel p is an ordinary homotopy equivalence and a pro $-\pi_*$ isomorphism at ∞ (ε(Tel p)$:\varepsilon$(Tel E) $\longrightarrow \varepsilon$(Tel B) is given by $\{p_n\}:\{E_n\} \longrightarrow \{B_n\}$ up to homotopy where ε is the ends functor) but Tel p is not a proper homotopy equivalence (at ∞). □

(5.5.11) Example [E -H -4]. We will construct below an inverse system X of simplicial sets such that pro $-\pi_*(X) = 0$, but X is not contractible. We do not know whether X can be chosen to be an inverse system of finite complexes or even the end of a locally finite complex.

Let $K(Z_2,n)$ denote the simplicial Eilenberg - MacLane space (see [May]). The direct system

$$K(Z_2,1) \xrightarrow{Sq^1} K(Z_2,2) \xrightarrow{Sq^2} K(Z_2,4) \xrightarrow{Sq^4} \cdots$$

has the property that all composite maps

$$\phi = Sq^{2^{n+k-1}} \cdots Sq^{2^{n+1}} Sq^{2^n} : K(Z_2,2^n) \longrightarrow K(Z_2,2^{n+k})$$

are essential (evaluate ϕ on the class x^{2^n} in $H^{2^n}(K(Z_2,1);Z_2)$ where x is the generator of $H^1(K(Z_2,1);Z_2)$; see N. E. Steenrod and D. B. A. Epstein [S - E]) but each bonding map kills π_* .

Form the inverse system X shown below.

$$X = \left\{ \begin{array}{ccccccc} \vdots & & \vdots & & \vdots & & \vdots \\ & & & \searrow & & \searrow & \\ X_3 & = & K(Z_2,1) & \times & K(Z_2,2) & \times & K(Z_2,4) \times \cdots \\ & & & \searrow & & \searrow & \\ X_2 & = & K(Z_2,.) & \times & K(Z_2,2) & \times & K(Z_2,4) \times \cdots \\ & & & \searrow & & \searrow & \\ X_1 & = & K(Z_2,1) & \times & K(Z_2,2) & \times & K(Z_2,4) \times \cdots \end{array} \right.$$

Then X has the required properties.

(5.5.12) Example. If $\pi = \{\pi_\alpha\}$ is a pro-group, we shall call $K(\pi,1) \equiv \{K(\pi_\alpha,1)\}$ the standard Eilenberg-MacLane pro-space with fundamental pro-group π. An Eilenberg-MacLane pro-space $X = \{X_n\}$ (i.e., $\{\tilde{X}_n\}$ is contractible) such that X is not equivalent to $\{K(\pi_1(X_n),1)\}$ in pro-Ho(SS_0) will be called exotic. There exist exotic Eilenberg-MacLane pro-spaces! Below we shall construct an exotic $K(Z_2,1)$ by (roughly) using Example (5.5.11) as the fibre of a fibration over the standard $K(Z_2,1)$, and twisting the bonding maps as in Example (5.5.10). If $K(\pi,1)$ is not finitely dominated, then there does not exist an inverse system of finite complexes X such that pro-$\pi_1(X) = \pi$ and $\tilde{X}$ is contractible. Whether such an exotic X can come from the end of a locally finite complex is an important question in infinite dimensional topology (see §7).

Let $F_n = \prod_{m=1}^{\infty} K(Z_2,2^m)$ for all n, let $X_n = F_n \times K(Z_2,1)$ for all n, and define twisted bonding maps $X_n \longrightarrow X_{n-1}$ so that the diagrams

$$\begin{array}{ccc} X_n & \longrightarrow & X_{n-1} \\ \downarrow & & \downarrow \\ K(Z_2,1) & \xrightarrow{id} & K(Z_2,1), \end{array} \qquad \begin{array}{ccc} X_n & \longrightarrow & X_{n-1} \\ \downarrow & & \downarrow \\ pt. & \longrightarrow & K(Z_2,2) \end{array}$$

and

(5.5.13)
$$\begin{array}{ccc} X_n & \longrightarrow & X_{n-1} \\ \downarrow & & \downarrow \\ K(Z_2,2^m) \times K(Z_2,1) & \xrightarrow{y \times x^{2^m}} & K(Z_2,2^{m+1}) \end{array}$$

$(m \geq 1)$

commute. In diagram (5.5.13), x in $H^1(K(Z_2,1);Z_2)$ and y in $H^{2^m}(K(Z_2,2^m;Z_2)$ are the generators, and the map $y \times x^{2^m}$ represents the cohomology class $y \times x^{2^m}$ in $H^{2^{m+1}}(K(Z_2,2^m) \times K(Z_2,1);Z_2)$.

As in Example (5.5.10), we obtain an inverse system of fibrations

(5.5.14)
$$\begin{array}{ccccccc} \cdots \longrightarrow & X_n & \xrightarrow{bond} & X_{n-1} & \longrightarrow \cdots \\ & \downarrow p_n & & \downarrow p_{n-1} & \\ \cdots \longrightarrow & K(Z_2,1) & \xrightarrow{id} & K(Z_2,1) & \longrightarrow \cdots . \end{array}$$

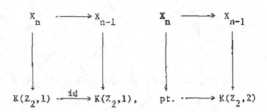

(5.5.15) <u>Theorem</u> [E-H-4]. The tower $X = \{X_n\}$ constructed above is an exotic Eilenberg-Maclane pro-space.

We sketch the proof for completeness. By construction the fibrations p_n in diagram (5.5.14) induce isomorphisms on π_1, so that $\tilde{X}_n \cong F_n$ for all n. Further, the bonding maps of $\{\tilde{X}_n\}$ and $\{F_n\}$ are compatible. By diagrams (5.5.13), the bonding maps of $\{F_n\}$ are null-homotopic. Hence $\{X_n\}$ is an Eilenberg-MacLane pro-space with fundamental pro-group Z_2.

If $\{X_n\}$ were isomorphic to the constant tower $K(Z_2,n)$. then the map $\{p_n\}$ would be an isomorphism. This would yield homotopy commutative diagrams

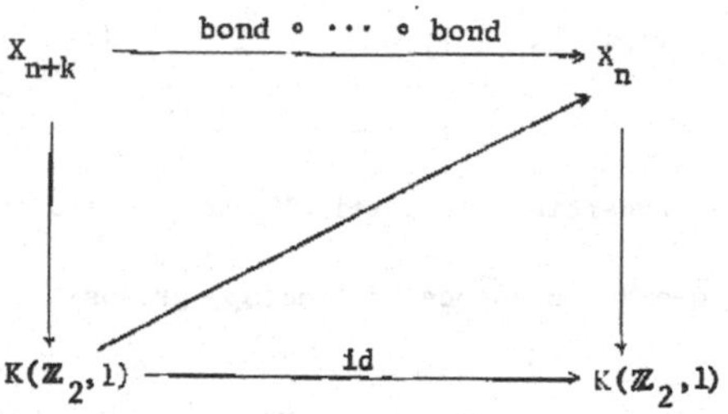

for each n and for suitable k depending upon n. Thus the induced map on cohomology $H^*(X_n,Z_2) \longrightarrow H^*(X_{n+k},Z_2)$ factors through $H^*(K(Z_2,1),Z_2)$, so that class $y \in H^*(X_{n+k},Z_2)$ maps to 0 in $H^*(X_n;Z_2)$. This contradicts the construction in diagram (5.5.13). Hence, $X = \{X_n\}$ is exotic. □

(5.5.16) <u>Example</u>. Let $X_n = \prod_{i=1}^{n} S^i$. Then $\text{pro-}\pi_i\{X_n\}$ is stable for all $i \geq 1$. Hence, the map $h:\text{holim}\,\{X_n\} \longrightarrow \{X_n\}$ is a ϕ-isomorphism.

Also, $X = \{X_n\}$ is movable. But X is not stable. Dydak [Dyd - 1] has shown that if X is movable and $\text{pro} - \pi_*(X)$ is stable, then X is stable.

§5.6. Strong homotopy and homology theories.

We shall define strong homotopy and homology theories on $\text{pro} - SS_*$. At present our main application is the development of generalized Steenrod homology theories on compact metric spaces (see §8).

(5.6.1) Definition. The strong homotopy groups of a pro - (pointed simplicial set) $\{X_j\}$ are given by

$$\pi_i\{X_j\} \equiv \text{Ho}(\text{pro} - SS_*)(S^i, \{X_j\}) = \text{Ho}(SS_*)(S^i, \text{holim}\ \{X_j\})$$

$$\equiv \pi_i(\text{holim}\ \{X_j\}).$$

These satisfy the usual properties, in particular a fibration sequence in $\text{pro} - SS_*$ yields a long-exact-sequence of homotopy groups.

(5.6.2) Strong stable homotopy groups. For any pro - (simplicial spectrum) $\{X_j\}$, define

$$\pi_i^s\{X_j\} \equiv \text{Ho}(\text{pro} - \text{Sp})(S^i, \{X_j\}) \cong \text{Ho}(\text{Sp})(S^i, \text{holim}\ \{X_j\})$$

$$\equiv \pi_i^s(\text{holim}\ \{X_j\}).$$

(5.6.3) Proposition. $\{\pi_i^s\}$ forms a generalized homology theory on pro - Sp.

Proof. Clearly the functors $\pi_i{}^s$ on pro - Sp are homotopy invariant. We begin by verifying the exactness axiom. Because each cofibration $A \to X$ is isomorphic to a levelwise cofibration $\{A_j\} \to \{X_j\}$ (Proposition (3.3.36)) it suffices to show that an inverse system of cofibration sequences $\{A_j \to X_j \to X_j/A_j\}$ induces three-term exact sequences $\pi_i{}^s\{A_j\} \to \pi_i{}^s\{X_j\} \to \pi_i{}^s\{X_j/A_j\}$. We may functorially replace $\{X_j/A_j\}$ by a fibrant pro - (simplicial spectrum) $\{Y_j\}$ $(X_j/A_j \xrightarrow{\simeq} Y_j)$, replace the map $\{X_j\} \to \{Y_j\}$ by a fibration $\{X'_j\} \to \{Y_j\}$ $(X_j \xrightarrow{\simeq} X'_j)$ in pro - Sp with (levelwise) fibre $\{F_j\}$, and form compatible commutative diagrams

(5.6.4)
$$\begin{array}{ccccc} A_j & \longrightarrow & X_j & \longrightarrow & X_j/A_j \\ \downarrow \simeq & & \downarrow \simeq & & \downarrow \simeq \\ F_j & \longrightarrow & X'_j & \longrightarrow & Y_j \end{array}$$

(The left weak equivalences arises because $\pi_i{}^s$ is exact on both cofibration and fibration sequences in Ho(Sp); loosely, cofibration and fibration sequences are "the same" in Ho(Sp).) Therefore the sequences

$\pi_i{}^s\{A_j\} \to \pi_i{}^s\{X_j\} \to \pi_i{}^s\{X_j/A_j\}$ and $\pi_i{}^s\{F_j\} \to \pi_i{}^s\{X'_j\} \to \pi_i{}^s\{Y_j\}$ are isomorphic. But the latter sequence is isomorphic to the sequence $\pi_i{}^s(\text{holim}\ \{F_j\}) \to \pi_i{}^s(\text{holim}\ \{X'_j\}) \to \pi_i{}^s(\text{holim}\ \{Y_j\})$, which is exact because an inverse system of fibrations is a fibration sequence in Ho(pro - Sp), holim $\simeq$ lim on the fibrant objects $\{F_j\}$, $\{X'_j\}$ and $\{Y_j\}$ (§4.2) and lim preserves fibration sequences (use Theorem (3.3.4)). The exactness axiom follows.

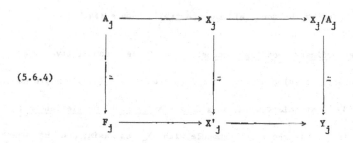

We may iterate this process to obtain a long-exact sequence.

For the suspension axiom, consider a cofibration sequence of the form $\{X_j\} \longrightarrow \{CX_j\} \longrightarrow \{\Sigma X_j\}$. By regarding this sequence as a fibration sequence, we obtain an exact sequence

$$0 = \pi_{i+1}{}^{s}\{CX_j\} \longrightarrow \pi_{i+1}{}^{s}\{\Sigma X_j\} \xrightarrow{\partial} \pi_i{}^{s}\{X_j\} \longrightarrow \pi_i{}^{s}\{CX_j\} = 0;$$

hence the suspension axiom holds. The conclusion follows. □

(5.6.5) Strong homotopy groups and homotopy pro-groups. Both unstably and stably these are related by the Bousfield-Kan spectral sequence (§4.9).

(5.6.6) Strong (ordinary) homology groups. Let R be a commutative ring with identity. We shall develop a strong homology theory ${}^{S}\tilde{H}_*(\ ;R)$ on pro-SS_*. Bousfield and Kan [B-K] associate with R a free simplicial R-module functor; for X in SS_*, $(RX)_n$ is the free R-module with X_n as basis, mod $R*$ where $*$ is the basepoint of X. There results a simplicial R-module, $RX = \{(RX)_n, d_i, s_i\}$, which depends functorially on X. Because R maps a cofibration sequence $A \longrightarrow X \longrightarrow X/A$ in SS_* into a fibration sequence $RA \longrightarrow RX \longrightarrow R(X/A)$, the functor $\pi_*(R-)$ is a (reduced) homology theory on SS_*. As with Dold and Thom's infinite symmetric product [D-T], $\pi_*(R-) \cong \tilde{H}_*(-;R)$.

We prolong the Bousfield-Kan functor R to pro-SS_* by defining $R\{X_j\} = \{RX_j\}$, and define

$${}^{S}\tilde{H}_*(\{X_j\};R) = \pi_*(R\{X_j\}).$$

Observe that R takes an inverse system of cofibration sequences into an inverse system of fibration sequences, which is a fibration sequence in $Ho(pro-SS_*)$ by Proposition (3.4.17). This yields the exactness axiom for ${}^S\tilde{H}_*(;R)$.

For the suspension axiom, first observe that natural map $\{X_j\} \longrightarrow R\{X_j\}$ (induced by the identity of R) yields maps $\{EX_j\} \longrightarrow \{ERX_j\} \longrightarrow \{\bar{W}RX_j\}$ where E is the simplicial suspension (see e.g. [May]). But $\{\bar{W}RX_j\}$ is a pro-(simplicial R-module) so we obtain a map

$$R\{EX_j\} \cong \{REX_j\} \longrightarrow \{\bar{W}RX_j\}.$$

It is easy to check that this map is a level weak equivalence. Hence

$${}^S\tilde{H}_*(\{X_j\};R) \cong \pi_*\{RX_j\}$$

$$\cong \pi_{*+1}\{\bar{W}RX_j\}$$

(use the fibration sequence

$$\{RX_j\} \longrightarrow \{WRX_j\} \longrightarrow \{\bar{W}RX_j\}$$

(3.4.17), and (5.6.1)).

Therefore, ${}^S\tilde{H}_*(-;R)$ is a homology theory on $pro-SS_*$. On SS_*, ${}^S\tilde{H}_*(-;R) \cong \tilde{H}_*(-;R)$.

(5.6.7) Strong (generalized) homology groups. Let $\tilde{h}_*$ be a generalized reduced homology theory on CW_*, which is represented by a CW spectrum E.

That is, $E = \{E_n | n \geq 0\}$, together with cellular inclusions $\Sigma E_n \longrightarrow E_{n+1}$, and

$$h_*(X) = \pi_*^s(X \wedge E) \equiv \pi_*^s\{X \wedge E_n\}$$

on CW_*. We prolong the smash product $- \wedge E$ to a functor from pro-SS_* to pro-Sp by defining

$$\{X_j\} \wedge E = \{\mathrm{Sin}((RX_j) \wedge E)\},$$

and set

$$^S\tilde{h}_*\{X_j\} = \pi_*^s(\{X_j\} \wedge E).$$

Because $- \wedge E$ takes cofibration sequences over Ho(pro-SS_*) into cofibration sequences over Ho(pro-Sp), and similarly, $- \wedge E$ preserves suspensions, $^S\tilde{h}_*\{X_j\}$ is a generalized reduced homology theory on pro-SS_*. As above, $^S\tilde{h}_* \equiv \tilde{h}_*$ on SS_*.

It is now easy to see that any map of spectra $E \longrightarrow F$ induces a natural transformation of strong generalized homology theories $^S\tilde{h}_* \longrightarrow {}^S\tilde{k}_*$ where E and F represent h_* and $\tilde{k}_*$ respectively. Thus all homology operations and products associated with h_* extend to $^S\tilde{h}_*$ (see [Adams-1, 3], also (2.2.65)).

(5.6.8) Proposition. For $\{X_j\}$ in SS_*, there is a Bousfield-Kan spectral sequence

$$E^2_{p,q} = \lim{}^p{}_j\{\tilde{h}_q(X_j)\},$$

which converges under suitable conditions to $^S\tilde{h}_*\{X_j\}$. If $\{X_j\}$ is isomorphic to a tower in $\mathrm{Ho}(\text{pro}-SS_*)$, the spectral sequence collapses to the short exact sequences

$$0 \longrightarrow \lim{}^1\{\tilde{h}_{n+1}(X_j)\} \longrightarrow {}^S\tilde{h}_n(X) \longrightarrow \lim\{\tilde{h}_n(X_j)\} \longrightarrow 0.$$

Proof. Use (5.6.6), (5.6.7) and the Bousfield-Kan spectral sequence (4.9.4). □

We conclude this section with several remarks about cohomology theories.

(5.6.9) Cohomology groups. If $\tilde{h}^*$ is a generalized, reduced cohomology theory on SS_*, then $\tilde{h}^*$ induces a cohomology theory on $\text{pro}-SS_*$:

$$\tilde{h}^*\{X_j\} \equiv \operatorname{colim}_j\{\tilde{h}^*(X_j)\},$$

because colim is exact. Further, if E_n represents $\tilde{h}^n$ on SS_*, then

$$\begin{aligned}\tilde{h}^*\{X_j\} &= \text{pro}-\mathrm{Ho}(SS_*)(\{X_j\},E_n)\\ &\cong \mathrm{Ho}(\text{pro}-SS_*)(\{X_j\},E_n).\end{aligned}$$

The latter isomorphism exists because E_n is stable.

(5.6.10) Representable theories. For $\{Y_k\} \in \text{pro}-SS_*$ there is an associated generalized cohomology theory defined by

$$\tilde{h}^n(-) = \mathrm{Ho}(\text{pro}-SS_*)(-,\Omega^{-n}\{Y_k\}),\ n \leq 0.$$

Conversely, Alex Heller [Hel-1, §11] showed that any group-valued cohomology theory on an h-c category (an abstraction of CW_* analogous to pointed closed model categories as abstractions of SS_*) is the colimit of a directed system of representable theories. Because Heller's proof only uses factorization through a stable category (such as pro - Sp), and properties of the homotopy relation and cofibrations, his result also holds for pro - SS_* and pro - Sp.

§6. PROPER HOMOTOPY THEORY

§6.1. Introduction.

In this section we shall use pro-homotopy theory to study the proper homotopy theory of locally compact, σ-compact Hausdorff spaces via a functor (the end) into pro-Top. This functor describes proper homotopy theory at ∞. We obtain proper homotopy theory by combining proper homotopy theory at ∞ with ordinary homotopy theory.

§6.2 contains the basic definitions of proper homotopy theory and the end functor.

In §6.3 we shall prove that the end functor from locally compact, σ-compact Hausdorff spaces to pro-Top yields a full embedding of the proper category at ∞.

We discuss proper Whitehead Theorems in §6.4. In particular we sketch a proof of L. Siebenmann's finite-dimensional proper Whitehead Theorem [Sieb-1] and show that an infinite-dimensional analogue (claimed by E. M. Brown [Br] and F. T. Farrell, L. R. Taylor, and J. B. Wagoner [F-T-W]) fails in general.

We discuss the Chapman complement theorem and an analogous "strong" complement theorem in §6.5.

§6.2. Proper homotopy and ends.

We shall show how T. Chapman's [Chap -1] formulation of proper homotopy theory leads to an embedding of the proper category at ∞ into a closed model category.

(6.2.1) Proper homotopy theory following Chapman. Call a continuous map $f:X \to Y$ of locally compact Hausdorff spaces proper if for each compactum $B \subset Y$ there is a compactum $A \subset X$ with $f(cl(X \setminus A)) \subset cl(Y \setminus B)$ (cl denotes closure). This is just a reformulation of the usual notion of proper map. Proper maps $f,g:X \to Y$ are called properly homotopic if there is a proper homotopy $H:X \times I \to Y$ with $H|_0 = f$ and $H|_1 = g$. Proper-homotopy-equivalences are now defined in the obvious way.

Chapman also introduced weak proper homotopy theory for the complement theorem. See §6.5.

(6.2.2) Definition. Let P be the category of locally compact Hausdorff spaces and proper maps, and $Ho(P)$ be its associated proper homotopy category. Let P_σ and $Ho(P_\sigma)$ be the restrictions to σ-compact spaces.

We shall study proper homotopy theory by associating to each proper map its ordinary homotopy class and its proper homotopy class at ∞. We describe proper homotopy theory at ∞ by introducing the following category of right fractions in the sense of P. Gabriel and M. Zisman [G - Z].

(6.2.3) The proper category at ∞. We shall call an inclusion $j:A \hookrightarrow X$ cofinal if the closure of the complement of A, $cl(X \setminus A)$, is compact; we shall also sometimes say that A is *cofinal* in X. Let Σ be the class of all cofinal inclusions in the proper category P. The quotient category

$$P_\infty \equiv P \setminus \Sigma$$

will be called the *proper category at* ∞. We shall sometimes call a morphism in P_∞ a *germ at* ∞ *of a proper map*.

It is easy to prove that $P \setminus \Sigma$ admits a calculus of right fractions. This means that each morphism from X to Y in $P \setminus \Sigma$ can be represented by a diagram

$$X \hookleftarrow A \xrightarrow{f} Y,$$

where A is cofinal in X and f is a proper map (morphism in P). Two such diagrams $X \hookleftarrow A' \xrightarrow{f'} Y$ and $X \hookleftarrow A'' \xrightarrow{f''} Y$ represent the same morphism if $f' = f''$ on a cofinal subspace of X. Composition is defined as follows. The class of cofinal inclusions Σ clearly contains identity maps and is closed under composition. Suppose that the diagrams $X \hookleftarrow A \xrightarrow{f} Y$ and $Y \hookleftarrow B \xrightarrow{g} Z$ represent germs at ∞ of proper maps. Form the solid-arrow diagram

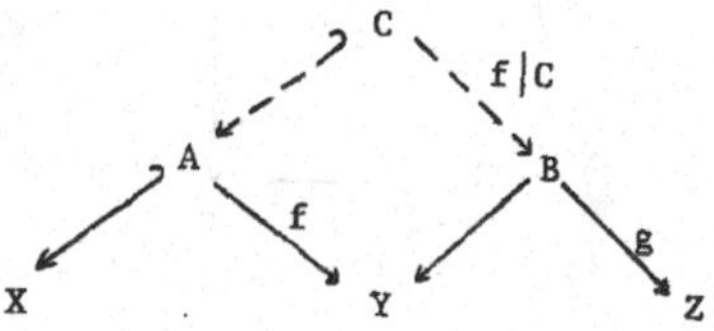

Let $C = f^{-1}(B)$. Because f is proper, C is cofinal in A, hence also cofinal in X. The required composite morphism is represented by

$$X \hookleftarrow C \xrightarrow{g \circ f|C} Z .$$

It is now easy to make the connection between proper homotopy theory at ∞ and pro- homotopy theory.

(6.2.4) <u>Definition</u>. Let X be a locally compact, Hausdorff space. The end of X is the inverse system

$$\varepsilon(X) = \{cl(X \setminus A) \mid A \text{ a compactum in } X\} ,$$

bonded by inclusion.

It is now easy to check that a map $f:X \longrightarrow Y$ of locally compact Hausdorff spaces is proper if and only if f induces a map $\varepsilon(f):\varepsilon(X) \longrightarrow \varepsilon(Y)$ which makes the diagram

$$\begin{array}{ccc} \varepsilon(X) & \xrightarrow{\varepsilon(f)} & \varepsilon(Y) \\ \downarrow & & \downarrow \\ X & \longrightarrow & Y \end{array}$$

commute. More generally, the following result holds, where $(1,\varepsilon)(X)$ is the pair $(X,\varepsilon(X))$.

(6.2.5) Proposition. The end construction yields the following functors and commutative diagrams:

$$\begin{array}{ccc} P_\infty & \xrightarrow{\ \varepsilon\ } & \text{pro-Top} \\ \downarrow & & \downarrow \\ \text{Ho}(P_\infty) & \xrightarrow{\ \varepsilon\ } & \text{Ho}(\text{pro-Top}), \end{array}$$

$$\begin{array}{ccc} P & \xrightarrow{\ (1,\varepsilon)\ } & (\text{Top, pro-Top}) \\ \downarrow & & \downarrow \\ \text{Ho}(P) & \xrightarrow{\ (1,\varepsilon)\ } & \text{Ho}(\text{Top, pro-Top}) \end{array}$$

The proof follows immediately from the definitions.

(6.2.6) Proposition. The functors $\varepsilon: P_\infty \longrightarrow$ pro - Top and $(1,\varepsilon): P \longrightarrow$ (Top, pro - Top) are full embeddings.

Again, this is clear from the definitions.

In the next section we shall prove the following.

(6.2.7) Proposition. The restrictions $\varepsilon: \text{Ho}(P_{\sigma,\infty}) \longrightarrow$ Ho(pro - Top) and $(1,\varepsilon): \text{Ho}(P_\sigma) \longrightarrow$ Ho(Top, pro - Top) are full embeddings.

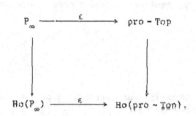

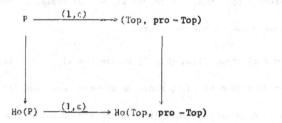

(6.2.8) Corollary. Let $f:X \to Y$ be a proper map of locally compact, σ-compact Hausdorff spaces. Then f is a proper homotopy equivalence if and only if f is an ordinary homotopy equivalence and a proper homotopy equivalence at ∞. □

(6.2.9) Remarks.

(a) The above results summarize our approach to proper homotopy theory. We use pro-homotopy to study proper homotopy theory at ∞, and then blend proper homotopy theory at ∞ and ordinary homotopy theory to obtain proper homotopy theory.

(b) The proof of Proposition (6.2.7) relies heavily on the telescope of a tower described in §3.7, i.e., on the coherent homotopy theory of towers. A suitable theory of coherent pro-homotopy should yield an extension of Proposition (6.2.7) and its corollary to all of P.

(6.2.10) Definitions. We shall call germs at ∞ of proper maps $f,g:X \rightrightarrows Y$ weakly properly homotopic at ∞ if the induced maps $\varepsilon(f)$, $\varepsilon(g)$ are equivalent in pro-Ho(Top). The corresponding weak proper homotopy category at ∞, $wHo(P_\infty)$ is then obtained by identifying maps in P_∞ which are weakly properly homotopic at ∞.

T. Chapman introduced the following weak proper homotopy category in order to prove the second part of the complement theorem [Chap-1], see §6.5. Call proper maps $f,g:X \rightrightarrows Y$ weakly properly homotopic if for each compactum B in Y

there is a compactum A in X and a homotopy $H: X \times I \longrightarrow Y$ (depending upon B) with $H|_0 = f$, $H|_1 = g$, and $H(\mathrm{cl}(X \setminus A) \times I) \subset \mathrm{cl}(Y \setminus B)$. Chapman's main concern was with contractible spaces. In this case, a straightforward application of Urysohn's lemma yields the following.

(6.2.11) Proposition. Let X and Y be contractible, locally compact, Hausdorff spaces. Then there is a bijection between weak-proper-homotopy classes of proper maps from X to Y and $\mathrm{wHo}(P_\infty)(X,Y)$.

Proof. There is clearly a functor from Chapman's category to $\mathrm{wHo}(P_\infty)$. For the converse, first let the diagram

$$X \hookleftarrow A \xrightarrow{f} Y$$

represent the germ of a proper map from X to Y. Because X is locally compact and $\mathrm{cl}(X \setminus A)$ is compact there is a compactum B in X with $\mathrm{cl}(X \setminus A) \subset \mathrm{int}\, B$. Because B is compact and Hausdorff, hence normal, there is a function $g: B \longrightarrow [0,1]$ with $g(\partial B) = 0$ and $g(\mathrm{cl}(X \setminus A)) = 1$. Choose a contracting homotopy for Y; $H : Y \times [0,1] \longrightarrow Y$ with $H|_0 = \mathrm{id}$ and $H|_1 = * \in Y$. We may define a proper map $\tilde{f}: X \longrightarrow Y$ by setting

$$\tilde{f}(x) = \begin{cases} f(x), & x \in X \setminus B \\ H(f(x), g(x)), & x \in A \cap B \\ *, & x \in X \setminus A \end{cases}$$

It is easy to check that $\tilde{f} = f$ in $\mathrm{wHo}(P_\infty)(X,Y)$ and that this construction yields a well-defined weak-proper-homotopy class of proper maps from X to Y. A similar argument shows that germs of proper maps which are equivalent in $wHo(P^{\bullet}_{\infty})$ yield the same weak-proper-homotopy class of maps. The conclusion follows. □

In the next section we shall prove that $\varepsilon: wHo(P_{\sigma,\infty}) \longrightarrow$ pro-Ho(Top) is a full embedding.

§6.3. Proper homotopy theory of σ-compact spaces.

We shall relate the proper homotopy theory of σ-compact spaces to the strong homotopy theory of towers, and thus prove Proposition (6.2.7).

(6.3.1) The end of a σ-compact space. Let X be a locally compact, σ-compact, Hausdorff space. Suppose that

$$X = \bigcup_{n=0}^{\infty} K_n$$

where $K_0 = \phi$, each K_n is compact, and each $K_n \subset \mathrm{int}(K_{n+1})$. Define $X_n = \mathrm{cl}(X \setminus K_n)$ for each n and let

$$\varepsilon'(X) = \{X_0 \supset X_1 \supset X_2 \supset \cdots\}.$$

Then $\varepsilon'(X)$ is a cofinal subtower of the end of X, $\varepsilon(X)$. Of course, any two cofinal subtowers of $\varepsilon(X)$ are canonically isomorphic in tow-Top. We may thus loosely regard ε' as a functor from P_σ to tow-Top and call $\varepsilon'(X)$ the end of X when there is no chance of confusion. Similarly, $(1,\varepsilon')$ may be loosely regarded as a functor from P_σ to (Top, tow-Top), or as a functor from P_σ to the category Filt of filterd spaces, see §3.7.

We shall associate to a space X in P_σ with end $\varepsilon'(X)$ (as in (6.3.1)) the <u>telescope</u> Tel $(\varepsilon(X))$, see §3.7, and <u>projection</u>

(6.3.2) $$p_X : \mathrm{Tel}\ (\varepsilon'(X)) \longrightarrow X, \quad p_X(x,t) = x.$$

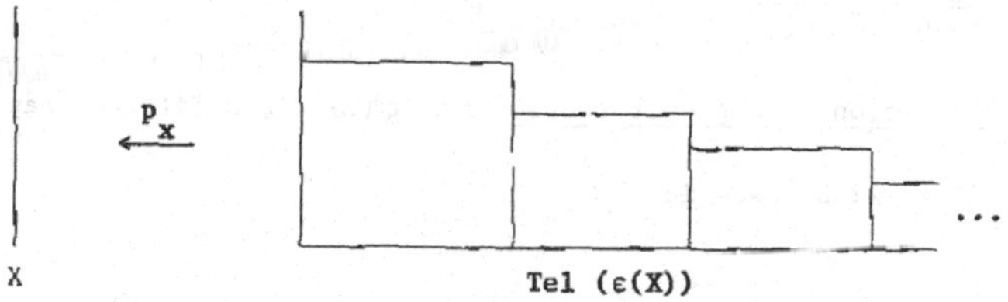

Then p_x is a filtered map (X is filtered by $\varepsilon'(X)$).

We shall need a suitable notion of naturality for Tel $(\varepsilon'(X))$ and p_X. Let $\varepsilon'(X)$ and $\varepsilon''(X)$ be two cofinal towers in the end of X. Because $\varepsilon'(X)$ and $\varepsilon''(X)$ are mutually cofinal, there is natural equivalence Tel $(\varepsilon'(X)) \simeq$ Tel $(\varepsilon''(X))$ in Ho(Tel) as in Proposition (3.7.13). Somewhat more is true. Maps $f_0, f_1 : W \rightrightarrows \mathrm{Tel}\ (\varepsilon'(X))$ with $p_X f = p_X f_1 = f : W \longrightarrow X$ are called <u>vertically homotopic</u> if there is a homotopy $H = \{H_t\} : W \times [0,1] \longrightarrow \mathrm{Tel}\ (\varepsilon(X))$ with $H_0 = f_0$, $H_1 = f_1$, and $pH_t = f$ for all t. We call H a <u>vertical homotopy</u>. If f_0, f_1, and H are also filtered maps, f_0 and f_1 are called <u>filtered-vertically-homotopic</u>. It is easy to prove the following.

(6.3.3) Lemma.

a) Tel $(\varepsilon'(X))$ and Tel $(\varepsilon''(X))$ are canonically equivalent up to filtered vertical homotopy.

b) The map p_X is natural in X.

Proof. For part (a) use the proof of Proposition (3.7.13). Part (b) follows immediately. □

(6.3.4) Definition. A proper section for $\varepsilon'(X)$ is a filtered map $s:X \longrightarrow \text{Tel}\,(\varepsilon'(X))$ with $ps = id_X$.

(6.3.5) Construction of proper sections [E - H - 5]. Urysohn's Lemma yields maps $h_n:cl(K_n \setminus K_{n-1}) \longrightarrow [n-2, n-1]$ with $h_n(K_{n-1}) = n-2$ and $h_n(\partial K_n) = n-1$ for $n \geq 2$. We may glue these maps together to obtain a proper map $h:X \longrightarrow R^+$ (notation: R^+ denotes the set of non-negative real numbers) such that $h(K_n \setminus K_{n-1}) \subset [n-2, n-1]$ for $n \geq 1$. Because $(X_{n-1} \setminus X_n) \subset \overline{(K_n \setminus K_{n-1})}$, there results a map $s:X \longrightarrow \text{Tel}\,(\varepsilon'(X))$, given by the formula

$$s(x) = (x, h(x)). \tag{6.3.6}$$

Clearly s is a proper section for $\varepsilon'(X)$. In fact, each proper section s' for $\varepsilon'(X)$ comes from a suitable proper map $h:X \longrightarrow R^+$ and formula (6.3.6).

(6.3.7) Proposition. X is a strong deformation retract of Tel $(\varepsilon'(X))$ in Filt.

<u>Proof</u>. The required retraction and inclusion are given by $p_X: \mathrm{Tel}\ (\varepsilon'(X)) \longrightarrow X$ and any proper section s for $\varepsilon'(X)$. The required homotopy from $id_{\mathrm{Tel}\ (\varepsilon'(X))}$ to sp_X is given by

$$H(x,t,t') = (x, (1-t')t + t' \cdot h(x)),$$

where $s(x) = (x, h(x))$. The arrows in the figure below represent H:

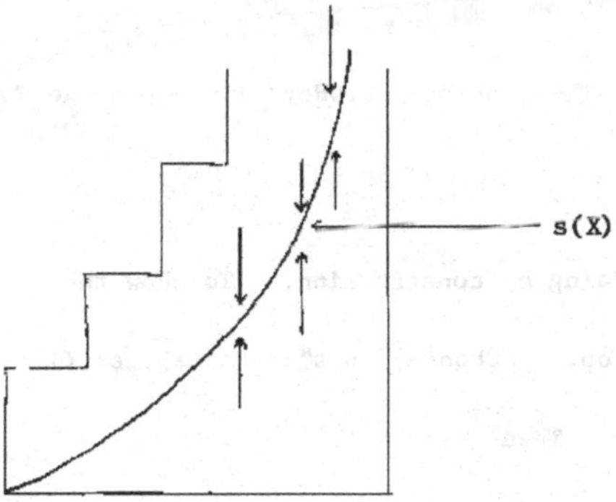

For each n choose $m > n$ so that $h(x_m) \subset [n,\infty)$. Then

$$H(\mathrm{Tel}\ (\varepsilon'(X))_m \times [0,1]) \subset \mathrm{Tel}\ (\varepsilon'(X))_n,$$

so that H is a filtered homotopy, as required. Note that H is even vertical. □

(6.3.8) <u>Proof of Proposition (6.2.7)</u>. We shall show that the functor $(1,\varepsilon): \mathrm{Ho}(P_\sigma) \longrightarrow \mathrm{Ho}(\mathrm{Top}, \mathrm{pro-Top})$ is a full embedding. The verification of the corresponding assertion about $\varepsilon: \mathrm{Ho}(P_{\sigma,\infty}) \longrightarrow \mathrm{Ho}(\mathrm{pro-Top})$ is easier and omitted. Let $X,Y \in P_\sigma$. Choose "ends" $\varepsilon'(X)$ and $\varepsilon'(Y)$ in tow-Top as in (6.3.1). By Proposition (6.3.7), proper homotopy classes of proper maps from X to Y are in bijective correspondence with filtered-homotopy classes of filtered maps from Tel $(\varepsilon'(X))$ to Tel $(\varepsilon'(Y))$. By Proposition (3.7.19), which states that filtered maps of telescopes yield a geometric model of Ho(Top, tow - Top), the latter class of maps is in bijective correspondence with

$$\text{Ho(Top, tow - Top)}(\varepsilon'(X),\varepsilon'(Y)) \cong \text{Ho(Top, pro - Top)}((X, \varepsilon(X)),(Y, \varepsilon(Y))),$$

see (6.3.1). The conclusion follows. □

(6.3.9) __Proposition.__ The functor $\varepsilon:\text{wHo}(P_\infty) \longrightarrow \text{pro - Ho(Top)}$ is a full embedding.

__Proof.__ ε is an embedding by construction. To show that ε is full, consider $\varepsilon(X)$, $\varepsilon(Y) \in$ pro - Top. Choose "ends" $\varepsilon'(X)$, $\varepsilon'(Y)$ cofinal in $\varepsilon(X)$, $\varepsilon(Y)$ as in (6.3.1). Then

$$\text{pro - Ho(Top)}(\varepsilon(X),\varepsilon(Y)) = \text{tow - Ho(Top)}(\varepsilon'(X),\varepsilon'(Y)),$$

$$\text{Ho(pro - Top)}(\varepsilon(X),\varepsilon(Y)) = \text{Ho(tow - Top)}(\varepsilon'(X),\varepsilon'(Y)).$$

Because the functor Ho(tow - Top) $\longrightarrow$ tow - Ho(Top) is surjective on maps (see (5.2.3)), we may realize any map in tow - Ho(Top) from $\varepsilon'(X)$ to $\varepsilon'(Y)$ by a map in Ho(tow - Top). The conclusion follows from Proposition (6.2.7) proved above. □

(6.3.10) __Proposition.__ The natural functor $\text{Ho}(P_{\sigma,\infty}) \longrightarrow \text{wHo}(P_{\sigma,\infty})$ induces a bijection on isomorphism classes of objects.

Proof. Use the above propositions and the corresponding result for towers in pro-homotopy (Corollary (5.2.17)). □

The above result is close to a "Whitehead-type" Theorem in proper homotopy theory; see §6.4.

§6.4. Whitehead-type Theorems.

L. Siebenmann [Sieb] gave various convenient criteria for a proper map of finite-dimensional, one-ended (see (6.4.1)) locally finite simplicial complexes to be a proper homotopy equivalence. Siebenmann's criterion implicitly involves "homotopy groups at ∞." In §5.5 we showed that Siebenmann's criterion fails without a finite-dimensional restriction (see (5.5.10d)). We shall discuss here three positive results: Siebenmann's; a result involving movability; and a result involving weak equivalences at ∞.

(6.4.1) Basepoints. We shall need basepoints in order to obtain the pro-homotopy groups of the end of locally compact space. We introduce "basepoints" as follows. Let X be a non-compact space in P (if X were compact, $\varepsilon(X) = \phi$, so basepoints for $\varepsilon(X)$ are irrelevant). Let $\omega: [0,\infty) \longrightarrow X$ be a proper embedding. We associate to X and ω the inverse system

$$\varepsilon(X,\omega) = \{(c\ell(X \setminus A) \cup \omega[0,\infty), \omega(0)) \mid A \text{ a compactum in } X\}$$

in pro-Top_*. Note that $\varepsilon(X,\omega) \cong \varepsilon(X)$ in Ho(pro-Top).

Call X one-ended if there is a unique proper homotopy class of proper maps

$[0,\infty) \longrightarrow X$ in $Ho(P)$, (equivalently, in $Ho(P_\infty)$).

(6.4.2) Homotopy "groups" at ∞. These are the pro-groups

$$\text{pro}-\pi_i(\varepsilon(X,\omega)), \quad i = 1,2,\cdots, \quad \text{and}$$

$$\text{pro}-\pi_*(\varepsilon(X,\omega)),$$

where

$$\pi_*(-) \equiv \prod_{i=1}^{\infty} \pi_i(-) \equiv [\bigvee_{i=1}^{\infty} S^i,-];$$

see §5.5.

In (5.5.10d) we constructed a proper map p of infinite-dimensional, one-ended countable, locally finite simplicial complexes which was an ordinary homotopy equivalence and induced an isomorphism on pro $-\pi_*$, but was <u>not</u> a proper homotopy equivalence.

However, by introducing suitable dimension restrictions or movability assumptions, one obtains the following positive results.

(6.4.5) <u>Theorem</u>. Let $f: X \longrightarrow Y$ be a proper map of one-ended, connected, countable, locally finite simplicial complexes which is an ordinary homotopy equivalence and induces isomorphisms $\text{pro}-\pi_*\varepsilon(X) \longrightarrow \text{pro}-\pi_*\varepsilon(Y)$. Then f is a proper homotopy equivalence if either of the following additional conditions holds.

a) [Sieb] $\dim X < \infty$ and $\dim Y < \infty$;

b) f is movable.

<u>Proof.</u> By Corollary (6.2.8), it suffices to verify that the induced map $\varepsilon(f)$ is invertible in Ho(pro - Top). Because X and Y are countable, $\varepsilon(X)$ and $\varepsilon(Y)$ admit cofinal subtowers $\varepsilon'(X)$ and $\varepsilon'(Y)$. Now use Theorem (5.5.6), with basepoints defined as above. □

We now describe a useful substitute for a true Whitehead Theorem. In their work on compactifying Q - manifolds, Chapman and Siebenmann [C - S] asked the following questions.

1) Is every weak-proper-homotopy equivalence a proper homotopy equivalence?

2) Is every weak-proper-homotopy equivalence weakly-properly-homotopic to a proper homotopy equivalence?

Chapman and Siebenmann confirmed a special case of (2), namely the case of Q - manifolds with tame ends. We obtained the general case in [E - H - 3]. It is much easier to obtain the following result, which in fact suffices for the applications in [C - S].

(6.4.6) <u>Theorem.</u> Let $f:X \longrightarrow Y$ be a map in P_σ such that $\varepsilon(f)$ is invertible in pro - Ho(Top) and f is invertible in Ho(Top). Then there is a proper homotopy equivalence $g:X \longrightarrow Y$ such that $\varepsilon(g) \simeq \varepsilon(f)$ in pro - Ho(Top) and $g \simeq f$ in Ho(Top).

<u>Proof.</u> By Theorem (5.2.9) there is a map $g':\varepsilon(X) \longrightarrow \varepsilon(Y)$ in pro - Top such that $g' \simeq \varepsilon(f)$ in pro - Ho(Top) and g' is invertible in Ho(pro - Top). Use Theorem (6.3.8) to realize g' as a map $g'':\overline{(X \setminus K_0)} \longrightarrow Y$ for some

compactum $K_0 \subset X$; i.e., $\varepsilon(g'') = g'$. For suitable K_0, $g'' \simeq f|(X \setminus K_0)$ in Ho(Top). Let $H:\overline{(X \setminus K_0)} \times [0,1] \longrightarrow Y$ be a homotopy from f to g''. Then choose a compactum $K_1 \subset X$ with $K_0 \subset \text{int } K_1$ and Urysohn function $h:K_1 \longrightarrow [0,1]$ with $h(K_0) = 0$ and $h(\text{bd } K_1) = 1$. The required map g is given by

$$g(x) = \begin{cases} f(x), & \text{for } x \in K_0, \\ H(x,h(x)), & \text{for } x \in K_1 \setminus K_0, \\ g''(x), & \text{for } x \in X \setminus K_1. \end{cases}$$

The required properties are easily verified. □

The following chart summarizes our use of pro-homotopy theory.

	Proper homotopy theory at ∞	Pro-homotopy theory
Strong	$Ho(P_\infty)$	Ho(pro-Top)
Weak	$wHo(P_\infty)$	pro-Ho(Top)

We shall see further connections in the next section.

§6.5 <u>The Chapman Complement Theorem</u>.

In the late 1960's Borsuk sparked an avalanche of interest in the study of the global homotopy properties of compacta (see [Mar-3], [Ed-1] and [E-H-2] for surveys of shape theory). Borsuk's original formulation of the shape theory of compact subsets of Hilbert space [Bor - 2] lacks the flexibility of the approach to be described in §8.2, but it has the advantage of being more geometric. This added geometry was quickly capitalized upon by Chapman in [Chap - 1].

Let $s = \prod_{n=1}^{\infty} (-1/n, 1/n)$ be the psuedo-interior of the Hilbert cube $Q = \prod_{n=1}^{\infty} [-1/n, 1/n]$. One defines the fundamental category or shape category, Sh, as follows. The objects of Sh are compact subsets of s. If X and Y are compact subsets of s, then a fundamental sequence $\underline{f}: X \to Y$ is defined as a sequence of maps $f_n: Q \to Q$ with the property that for every neighborhood V of Y in Q there exists a neighborhood U of X in Q and an integer n_0 such that for $n, n' \geq n_0$ the restrictions $f_n|U$ and $f'_n|U$ are homotopic in V. Note that $f_n(X)$ does not have to be contained in Y; it only has to be near Y. Two fundamental sequences $\underline{f}, \underline{f}': X \rightrightarrows Y$ are considered homotopic, $\underline{f} \simeq \underline{f}'$, provided that for every neighborhood V of Y in Q there exists a neighborhood U of X in Q and an integer n_0 such that for $n \geq n_0$, $f_n|U$ and $f_{n'}|U$ are homotopic in V. The morphisms in Sh are now taken to be homotopy equivalence classes of fundamental sequences. Two compacta X and Y contained in s are said to have the same shape if they are isomorphic in Sh.

In [Chap - 1] Chapman proved the following beautiful theorem.

(6.5.1) Chapman Complement Theorem. If X and Y are compacta in s, then X and Y have the same shape if and only if their complements $Q \setminus X$ and $Q \setminus Y$ are homeomorphic.

Chapman then extended the association $X \mapsto Q \setminus X$ to a functor T from the shape category to the weak proper homotopy category of complements in Q of compacta in s.

(6.5.2) Definition. Let $P_{Q,\infty}$ be the category of complements in Q of compacta in s and germs at ∞ of proper maps.

Because Q is contractible, Chapman's weak proper homotopy category above is isomorphic to $wHo(P_{Q,\infty})$ (use Proposition (6.2.11)).

Chapman then proved the following categorical version of the Complement Theorem, stated in our language.

(6.5.3) Theorem. There is a category isomorphism $T: Sh \longrightarrow wHo(P_{Q,\infty})$.

Outline of proof. Let K and L be compacta in s. Let $U = \{U_j\}$ and $V = \{V_k\}$ be bases of open neighborhoods for K and L, respectively, in Q. Chapman showed that K and L are Z-sets, hence the natural inclusions

$$\{U_j \setminus K\} \hookrightarrow \{U_j\}$$

and

$$\{V_k \setminus L\} \hookrightarrow \{V_k\}$$

are levelwise homotopy equivalences. It is easy to show that $Sh(X,Y) \cong \text{pro-}Ho(Top)(U,V)$. We then obtain a string of isomorphisms

$$
\begin{aligned}
Sh(X,Y) &\cong \text{pro-}Ho(Top)(U,V) \\
&\cong \text{pro-}Ho(Top)(\{U_j \setminus K\},\{V_k \setminus L\}) \\
&\cong wHo(P_{Q,\infty})(Q \setminus K,\ Q \setminus L)
\end{aligned}
$$

(the complements of V_j and U_k are compact in Q). The conclusion follows.

(6.5.4) <u>The strong shape category</u>. Let K be a compactum in s. It is easy to check that the complement $Q \setminus K$ is σ-compact, and hence that K has a countable basis of open sets U_j with $cl(U_j) \subset U_{j-1}$ $(j = 0,1,2,\cdots)$. We may therefore <u>define</u> the <u>strong shape category</u> s-Sh to be the category of compacta in s and coherent homotopy classes of maps of their associated neighborhood bases in Q; namely

$$
\begin{aligned}
(6.5.4) \qquad s\text{-}Sh(K,L) &\equiv Ho(Tel)(\text{Con Tel }\{U_j\},\ \text{Con Tel }\{V_k\}) \\
&\cong Ho(Tel)(\text{Con Tel }\{cl(U_j)\},\ \text{Con Tel }\{cl(V_k)\})
\end{aligned}
$$

where Con Tel is the contractible telescope

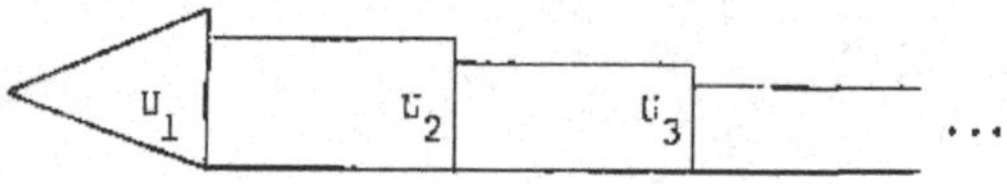

see §3.7. By Proposition (3.7.20),

$$(6.5.5) \qquad s\text{-}Sh(K,L) \cong Ho(\text{pro-}Top)(\{U_j\},\{V_k\}).$$

By following the proof of Theorem (6.5.3), we obtain the following strong categorical version of the Complement Theorem.

(6.5.6) <u>Theorem</u>. There is a commutative diagram of categories and functors

$$\begin{array}{ccc} s-Sh(K,L) & \xrightarrow{\cong} & Ho(P_{Q,\infty}) \\ \downarrow & & \downarrow \\ Sh(K,L) & \xrightarrow{\cong} & wHo(P_{Q,\infty}) . \end{array} \qquad \square$$

(6.5.7) <u>Remarks</u>. With reference to the original Chapman Complement Theorem (6.5.1), recall that the functor $Ho(P_{Q,\infty}) \longrightarrow wHo(P_{Q,\infty})$ induces a bijection on isomorphism classes of objects (Proposition (6.3.10)).

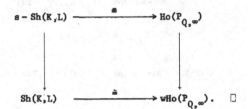

§7. GROUP ACTIONS ON INFINITE DIMENSIONAL MANIFOLDS.*

§7.1. Introduction.

We shall discuss the classification of actions of compact Lie groups on

$s = \prod_{n=1}^{\infty} (-1/n, 1/n)$ and on $Q = \prod_{n=1}^{\infty} [-1/n, 1/n]$.

In §7.2 we shall review the theory of s-manifolds and Q-manifolds.

Standard group actions will be constructed in §7.3, following Jim West [West-1]. We shall also show that all principal actions on s are standard up to equivariant homeomorphism.

§7.4 contains a classification theorem (largely due to West) for semifree actions of a finite group on Q. Whether such actions are unique is an open, interesting, and deep question.

§7.2. Basic theory of s-manifolds and Q-manifolds.

An s-manifold (respectively, Q-manifold) is a separable metric space locally homeomorphic to an open set in s (respectively, Q).

(7.2.1) s-manifolds. Work of Kadec, Bessaga, and Pelczynski culminated in Anderson's proof that all separable Frechet spaces are homeomorphic (see [A-B]).

The classification theory of s-manifolds is due to Henderson [Hend]. First,

*This section represents joint work with Jim West and much of the material is taken from his unpublished notes [West-1].

every s - manifold M can be triangulated, that is, there is a locally finite simplicial complex K such that M is homeomorphic to $K \times s$. Second, every homotopy equivalence $f:M_1 \longrightarrow M_2$ between s - manifolds is homotopic to a homeomorphism.

Torunczyk [Tow - 1,2] has shown that $X \times s$ is an s - manifold if and only if X is an ANR (separable, metric).

(7.2.2) Q - manifolds. Many of the above results hold for Q - manifolds if homotopy equivalence is replaced by ∞ - simple homotopy equivalence (see §7.2.3). Keller [Kel] showed that Q is homogeneous.

The classification theory of Q - manifolds is due to Chapman [Chap - 1 - 7] and West [West - 2,3]. West [West - 2] showed that $K \times Q$ is a Q - manifold if K is a locally finite simplicial complex. Chapman [Chap - 4] proved the converse: every Q manifold is triangulable, i.e., homeomorphic to $K \times Q$ for some locally finite simplicial complex K. If a map $f:K \longrightarrow L$ of locally finite simplicial complexes is an ∞ - simple homotopy equivalence, then $f \times id_Q:K \times Q \longrightarrow L \times Q$ is properly homotopic to a homeomorphism [West - 2]. The converse also holds [Chap - 3] and implies the topological invariance of Whitehead torsion.

West [West - 3] showed that every locally compact ANR is the CE - image of a Q - manifold; thus every compact ANR has finite homotopy type. Chapman [Chap - 7] observed that West's result can be used to extend ∞ - simple homotopy theory to locally compact ANR's; in particular, every such ANR has the ∞ - simple homotopy type of a locally finite simplicial complex. Hence, a proper map $f:M_1 \longrightarrow M_2$

between Q-manifolds is properly homotopic to a homeomorphism if and only if f is ∞-simple.

Bob Edwards [Edw] has recently shown that $X \times Q$ is a Q-manifold if (and only if) X is a locally compact metric ANR.

We shall conclude this section by giving West's [West-1] treatment of "infinite simple homotopy theory after Siebenmann."*

(7.2.3) The basic idea of (finite) simple homotopy theory [Cohen], is to single-out and study maps of finite cell complexes which are homotopic to finite compositions of maps which are of the form of an inclusion $i: L \longrightarrow K$, where $K = L \cup \mathring{e}^{n-1} \cup \mathring{e}^{n}$ with e^{n-1} an n-cell which is a face of e^n, or are of the form of a homotopy inverse to such an inclusion. In generalizing this notion to locally finite and not necessarily finite-dimensional complexes, one replaces the map i above by an inclusion $j: L \longrightarrow K$ in which $\overline{K \setminus L}$ is the union $\cup K_i$ of disjoint complexes K_i each of which collapses to $\overline{K}_i \cap L$, i.e., $(K_i \cap L) \longrightarrow K_i$ is the result of a finite sequence of inclusions such as i, and one restricts oneself to proper mappings and proper homotopies.

(7.2.4) $\mathcal{S}(K)$. In his treatment [Sieb] of infinite simple homotopy theory, Siebenmann introduces the group $\mathcal{S}(K)$ of simple structures on K. Each element of $\mathcal{S}(K)$ is an equivalence class $[f: K \longrightarrow L]$ of proper homotopy equivalences with domain K, where $g: K \longrightarrow M$ is in the class of f whenever there is a

---
*This survey is a quote from [West-1].

simple homotopy equivalence $s:L \longrightarrow M$ such that $g = sf$. (The group operation on $S(K)$ need not concern us here, but it is (essentially) geometrically defined in [Sieb] by representing [f] and [g] with inclusions, say, into the tops (domains) of mapping cylinders, and then [f] [g] is represented by the inclusion of K into the result of identifying the two spaces along the copies of K.) In particular, if there is only one simple structure on K, then all proper homotopy equivalences are simple.

(7.2.5) <u>Wh</u>, <u>K_0</u>, <u>and limits.</u> To examine $S(K)$, Siebenmann uses the Whitehead functor Wh and the projective class group functor K_0 in several limiting constructions at the end of K. (In essence, all that is needed for this paper is that these are functors.) The limiting constructions are as follows. Let the end of K be

$$\varepsilon(K) = \{K = W_0 \supset W_1 \supset W_2 \supset \cdots\},$$

(see (6.2.14)), where the W_n are subcomplexes of K whose complements have compact closure. Choose a proper <u>base ray</u> $\omega:[0,\infty) \longrightarrow K$ as in §6.5. Let $\varepsilon(K,\omega) = \{(W_n \cup \omega[0,\infty),\omega(0))\}$. Now consider the inverse system

$$\pi_1\varepsilon(K) \equiv \pi_1\varepsilon(K_1,\omega) \equiv \{\pi_1(W_n, \omega(0))\}.$$

Now define

$$K_0\ \pi_1\varepsilon(K) \equiv \lim_n\{K_0\ \pi_1(W_n, \omega(0)),\ K_0(i_*)\},$$

$$\mathrm{Wh}\ \pi_1\varepsilon(K) \equiv \lim_n\{\mathrm{Wh}\ \pi_1(W_n, \omega(0)),\ \mathrm{Wh}(i_*)\},$$

and the <u>attenuation</u>

$$\mathrm{Wh}\ \pi_1\varepsilon'(K) \equiv \lim{}^1_n\{\mathrm{Wh}\ \pi_1(W_n, \omega(0)), \mathrm{Wh}(i_*)\}.$$

Observe that the attenuation is zero if $\pi_1\varepsilon(K)$ is stable, i.e., pro-isomorphic to a group.

<u>Exact sequences</u>. Siebenmann gives two exact sequences to aid in computing $S(k)$. They are as follows:

(7.2.6) $\quad 0 \longrightarrow S_b(K) \longrightarrow S(K) \longrightarrow K_0\ \pi_1\varepsilon(K) \longrightarrow K_0\ \pi_1(K);$

(7.2.7) $\quad \mathrm{Wh}\ \pi_1\varepsilon(K) \longrightarrow \mathrm{Wh}\ \pi_1(K) \longrightarrow S_b(K) \longrightarrow \mathrm{Wh}\ \pi_1\varepsilon'(K) \longrightarrow 0.$

($S_b(K)$ is the group of equivalence classes of proper homotopy equivalences defined analogously to $S(K)$ but where one allows the inclusion $K_i \cap L \longrightarrow K_i$ to be <u>any</u> inclusion of finite complexes which is a homotopy equivalence.)

§7.3. <u>The Standard Actions</u>.

Following West [West-1], we shall construct standard principal actions of any compact Lie group G on Q_0 (Q with a point deleted) and on s. We shall show that all principal actions of G on s are standard. Let G act on itself by left translation. This principal action extends to a semi-principal action with unique fixed point on the cone of G, $C(G) \equiv G \times [0,1] / G \times \{0\}$. The product action on $\prod_{i=1}^{\infty} C(G)$ is also semi-principal with unique fixed point, the infinite cone point. But, it follows from [West - 2] that $\prod_{i=1}^{\infty} C(G)$ is homeomorphic to Q. Removing the unique fixed point yields the <u>standard</u> principal action σ_G of G on Q_0. Since Q_0 is contractible, and $Q_0 \times s$ is an s-manifold, $Q_0 \times s$ is homeomorphic to s. Thus, we also obtain a standard action ρ_G of G on s. Any principal action of G on Q_0 or s which is not conjugate to the standard action will be called <u>exotic</u>.

Let ρ and ρ' be two principal actions of G on s. We will say that the actions are <u>nice</u> if the quotient spaces s/ρ and s/ρ' are s-manifolds (e.g., if G is finite, then this is always the case). In any case, s/ρ and s/ρ' are both classifying spaces for G. Let $f: s/\rho \longrightarrow s/\rho'$ be a map such that the induced bundle $f^*(s,\rho')$ over s/ρ is isomorphic to the bundle (s,ρ) (see [Hus]). Since f is a homotopy equivalence, if s/ρ and s/ρ' are both s-manifolds, then f is homotopic to a homeomorphism g. One thus obtains the diagram of principal G-bundle isomorphisms

(7.3.1)
$$\begin{array}{ccccc} s & \xrightarrow{\ \phi\ } & s & \xrightarrow{\ \psi\ } & s \\ \downarrow & & \downarrow & & \downarrow \\ s/\rho & \longrightarrow & s/\rho & \xrightarrow{\ g\ } & s/\rho' \end{array}$$

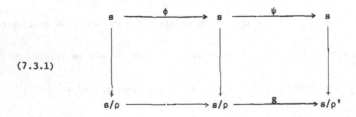

in which $h = \psi \circ \phi$ is a G-equivariant homeomorphism from (s,ρ) to (s,ρ'); hence ρ and ρ' are equivalent actions of G on s. Summarizing, we have the following theorems.

(7.3.2) Theorem. All nice principal actions of a compact Lie group G on s are standard. □

(7.3.3) Theorem. All free actions of a finite group G on s are standard. □

The Q_0 case is much more subtle, since one must show that Q_0/ρ and Q_0/ρ' have the same ∞-simple homotopy type, and not just the same homotopy type, before one can conclude that they are homeomorphic. The main result of this section is the following theorem, which will be proved in §7.4.

(7.3.4) Theorem. Let ρ and ρ' be free actions of a finite group G on Q_0. Then the following statements are equivalent:

1) ρ is equivalent to ρ';

2) Q_0/ρ is homeomorphic to Q_0/ρ';

3) Q_0/ρ is ∞-simple homotopy equivalent to Q_0/ρ';

4) Q_0/ρ is proper homotopy equivalent to Q_0/ρ';

5) The end of Q_0/ρ, $\varepsilon(Q_0/\rho)$ is homotopy equivalent to $\varepsilon(Q_0/\rho')$ in pro-Ho(Top).

(7.3.5) Remarks. The equivalence of (1) - (4) for G a finite group is due to West [West-1]. The end $\varepsilon(Q_0/\rho)$ is a quotient of $\varepsilon(Q_0) \simeq$ pt.; hence, $\varepsilon(Q_0/\rho)$ is a pro-space analog of the classifying space $BG = Q_0/\rho$. West showed that the natural inclusion $\varepsilon(Q_0/\rho) \longrightarrow Q_0/\rho$ is always a $\natural$-isomorphism;

even an isomorphism on pro $-\pi_*$. By Theorem 7.3.4, it is a homotopy equivalence if and only if ρ is standard. In §5 we showed the existence of uncountably many exotic $K(Z_2,1)$'s; but we still do not know of any exotic compact Lie group actions on Q_0. On the other hand, work of Tucker [Tuc - 2] shows that there are uncountably many different actions of Z on Q_0.

§7.4. Proof of Theorem (7.3.4).

The following preliminary lemmas, as well as the equivalence of statements (1)-(4) in (7.3.4) are taken from [West - 1]. Our machinery (pro - spaces) is used to simplify some of the statements and arguments.

Let G be a fixed finite group acting semifreely on Q with unique fixed point q. Let

$$\alpha: G \times Q \longrightarrow Q, \qquad (g,x) \overset{\alpha}{\longmapsto} gx$$

denote the action.

(7.4.1) Lemma. $Q \setminus \{q\}$ is contractible, and its end $\varepsilon(Q \setminus \{q\})$ is contractible (in Ho(pro - Top)).

Proof. Represent Q as the product $\prod_{i=1}^{\infty} [0,1]_i$. Because Q is homogeneous (see 7.2.2), we may assume that $q = (0,0,0,\cdots)$. $Q \setminus \{q\}$ is then convex, hence contractible. Also,

$$\varepsilon(Q \setminus \{q\}) \cong \{U_i \equiv (0,1/i]^i \times \prod_{j>i} (0,1]_j \mid i \geq 1\},$$

bonded by inclusion. Because each U_i is convex, hence contractible, $\varepsilon(Q \setminus \{q\})$

is contractible in pro-Ho(Top), hence in Ho(pro-Top) by Corollary (5.2.17). □

(7.4.2) <u>Lemma</u>. There is a commutative diagram of covering maps in pro-Top

$$(7.4.3)\qquad \begin{array}{ccc} G & = & G \\ \downarrow & & \downarrow \\ \varepsilon(Q\setminus\{q\}) & \longrightarrow & Q\setminus\{q\} \\ \downarrow & & \downarrow \\ \varepsilon((Q\setminus\{q\})/\alpha) & \longrightarrow & (Q\setminus\{q\})/\alpha. \end{array}$$

<u>Proof</u>. First, choose a representative tower for $\varepsilon(Q\setminus\{q\})$ as follows. Let $U_0 = Q\setminus\{q\}$ and $\{U_i \mid i \geq 1\}$ be as in (7.4.1). For each i, $\bigcap_{g\in G} gU_i$ (the intersection of translates of U_i under α) contracts in $U_i\setminus\{q\}$. Let

$$(7.4.4)\qquad \varepsilon(Q\setminus\{q\}) \equiv \{V_0 \supset V_1 \supset V_2 \supset \cdots\}$$

be a subsequence of $\{\bigcap_{g\in G} gU_i\}$ chosen so that V_i contracts in V_{i-1}. Note that each V_i is invariant under α.

Next, rewrite diagram (7.4.3) as

$$(7.4.5)\qquad \begin{array}{ccc} G & = & G \\ \downarrow & & \downarrow \\ \varepsilon(Q\setminus\{q\}) & \longrightarrow & Q\setminus\{q\} \\ \big\downarrow{\scriptstyle \varepsilon(p)} & & \big\downarrow{\scriptstyle p} \\ \varepsilon((Q\setminus\{q\})/\alpha) & \longrightarrow & (Q\setminus\{q\})/\alpha, \end{array}$$

where p is the covering map induced by α, and $\varepsilon(p)$ is the levelwise covering map

$$\varepsilon(Q \setminus \{q\}) = \{V_i\} \longrightarrow \{V_i/\alpha\} \equiv \varepsilon((Q \setminus \{q\})/\alpha),$$

yielding the conclusion. □

(7.4.6) Corollary. $\varepsilon((Q\setminus \{q\})/\alpha)$ is an Eilenberg-MacLane pro-space with pro-$\pi_1 = G$; i.e., the tower of universal covers $\{(V_i/\alpha)^{\sim}\}$ is contractible in Ho(pro-Top).

Proof. It remains only to observe that $\{(V_i/\alpha)^{\sim}\} \cong \varepsilon(Q \setminus \{q\})$; this is an easy exercise involving covering spaces. □

(7.4.8) Proof of Theorem (7.3.4). The following implications are easy:

(1) ⟺ (2) by covering space theory;

(2) ⟺ (3) by Chapman and West's classification of Q-manifolds (see (7.2.2));

(3) ⟹ (4) by definition; and

(4) ⟹ (5) by definition.

To verify (4) ⟹ (3), we shall show that for any semi-free action α of a finite group G on Q with unique fixed point q, $S((Q\setminus \{q\})/\alpha) = 0$, so that any proper homotopy equivalence with domain $(Q\setminus \{q\})/\alpha$ is ∞-simple. Triangulate $(Q\setminus \{q\})/\alpha$ as $K \times Q$ for some locally finite simplicial complex K. Because the projection map $(Q\setminus \{q\})/\alpha \longrightarrow K$ is a proper homotopy equivalence,

(7.4.9) $$\pi_1\varepsilon(K) \cong \pi_1 K \cong G,$$

via the inclusion. Hence, $K_0\,\pi_1\varepsilon(K) = K_0\,\pi_1(K)$, so that $S(K) \cong S_b(K)$ by (7.2.6). Also, $\mathrm{Wh}\,\pi_1\varepsilon(K) \cong \mathrm{Wh}\,\pi_1(K)$ and $\mathrm{Wh}\,\pi_1\varepsilon'(K) = 0$ (by (7.4.9), see (7.2.5)), so that $S_b(K) = 0$ by (7.2.7). Hence $S(K) = 0$, as required.

To verify (5) $\Rightarrow$ (4), first consider the diagram

$$\begin{array}{ccc} \operatorname{holim} \varepsilon(K) & \longrightarrow & \operatorname{holim} K \\ \downarrow & & \downarrow \\ \varepsilon(K) & \longrightarrow & K \end{array}$$

induced by (7.4.5) (the homotopy inverse limit, holim, is developed in §§4.1-4.2 and §4.9), where K is as above. By Corollary (7.4.6), $\text{pro}-\pi_i(\varepsilon(K)) = 0$ unless $i = 1$, in which case $\text{pro}-\pi_1(\varepsilon(K)) = G$. Applying the Bousfield-Kan spectral sequence to $\operatorname{holim} \varepsilon(K)$ yields

$$\pi_i \operatorname{holim} \varepsilon(K) = \begin{cases} G, & i = 1, \\ 0 & i \geq 1 \end{cases}$$

(note that $\operatorname{holim} \varepsilon(K)$ is pointed and connected). Hence the map $\operatorname{holim} \varepsilon(K) \longrightarrow K$ is a homotopy equivalence by the ordinary Whitehead theorem (K is a $K(G,1)$ by Lemmas (7.4.1) and (7.4.2)). Therefore K is a retract of $\varepsilon(K)$ in Ho(pro - Top) by the diagram

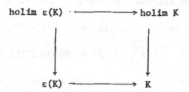

$$\begin{array}{ccc} \operatorname{holim} \varepsilon(K) & \xrightarrow{\ \simeq\ } & \operatorname{holim} K \\ \big\downarrow & \nwarrow & \big\downarrow \simeq \\ \varepsilon(K) & \longrightarrow & K \end{array} \quad .$$

[In the diagram above, the diagonal arrow from K to $\operatorname{holim} \varepsilon(K)$ is dashed.]

Now, for two semi-free actions ρ and ρ' of G on Q with unique fixed points q and q', assume $(Q \setminus \{q\})/\rho$ and $(Q \setminus \{q'\})/\rho'$ are proper homotopy equivalent at ∞. Then $\varepsilon((Q \setminus \{q\})/\rho) \cong \varepsilon((Q \setminus \{q'\})/\rho')$ in $\mathrm{Ho(pro-Top)}$ by Theorem (6.3.4). This equivalence extends to the diagram

(7.4.11)
$$\begin{array}{ccc} \varepsilon((Q \setminus \{q\})/\rho) & \rightleftarrows & (Q \setminus \{q\})/\rho \\ \big\downarrow \simeq & & \big\downarrow \simeq \\ \varepsilon((Q \setminus \{q'\})/\rho') & \rightleftarrows & (Q \setminus \{q'\})/\rho' \end{array}$$

[In diagram (7.4.11), the right-hand vertical arrow and the two leftward arrows are dashed.]

by (7.4.10). Theorem (6.3.3) now implies that $(Q \setminus \{q\})/\rho$ and $(Q \setminus \{q'\})/\rho'$ are proper homotopy equivalent (globally), as required. □

(7.4.12) <u>Remarks</u>. The implications (1) $\Longrightarrow$ (2) $\Longleftrightarrow$ (3) $\Longleftrightarrow$ (4) $\Longleftrightarrow$ (5) hold for arbitrary compact Lie groups G with a similar proof. For (2) $\Longrightarrow$ (1) we must verify that the maps $(Q \setminus \{q\})/\rho \longrightarrow (Q \setminus \{q'\})/\rho'$ of statement (2) are covered by equivariant maps $Q \setminus \{q\} \longrightarrow Q \setminus \{q'\}$.

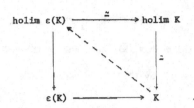

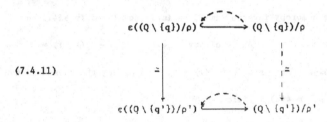

§8. STEENROD HOMOTOPY THEORY

§8.1. Introduction.

This chapter is concerned with shape theory, shape functors, and Steenrod homology theories.

In §8.2 we discuss the Steenrod homology theory [St - 1], the Kaminker-Schochet [K - S] axioms for generalized Steenrod homology theories, and the Vietoris construction [Por - 1]. We then use the Vietoris construction to discuss Steenrod homotopy theory. In particular, we define canonical Steenrod and Čech extensions, ${}^{S}h_*$ and $\check{h}_*$, of any generalized homology theory, h_*, defined on finite CW complexes. Proofs of the properties of ${}^{S}h_*$ occupy the next four sections.
D. S. Kahn, J. Kaminker, and C. Schochet have independently obtained Steenrod extensions by different methods.

In §8.3 we verify useful properties of the Vietoris functor.

In §8.4 we prove that ${}^{S}h_*$ is a homology theory on the category CM of compact metric spaces, that is, that ${}^{S}h_*$ is a homotopy invariant functor which satisfies the first two Kaminker-Schochet axioms for a Steenrod homology theory. We also show that products and operations associated with h_* extend to ${}^{S}h_*$.

We give two spectral sequences converging to ${}^{S}h_*$ in §8.5, and use the first spectral sequence to prove that ${}^{S}h_*$ satisfies the remaining Kaminker-Schochet axiom.

In §8.6 we prove functional and Steenrod duality theorems for ${}^{S}h_*$. Steenrod duality states that for a compactum X in S^n, ${}^{S}h_p(X) = h^{n-p-1}(S^n \setminus X)$. Functional duality is due to D. S. Kahn, J. Kaminker, and C. Schochet [K-K-S].

§8.2. Steenrod homology theories.

Let h_* be a generalized homology theory defined on the category of finite CW complexes. In this section we shall define a canonical Steenrod extension ${}^{S}h_*$ (see (8.2.3)) of h_* to the category CM of compact metric spaces. Proofs occupy §§8.3-8.5. See (8.2.1) and (8.2.2) for ordinary Steenrod homology and Steenrod K-homology.

The problem of constructing generalized Steenrod homology theories with "good properties" was posed by M. F. Atiyah and others at the Operator Theory and Topology Conference held at the University of Georgia in April, 1975. By "good properties" Atiyah meant that products and homology operations extended from h_* to ${}^{S}h_*$. Our extension enjoys these properties.

(8.2.1) Duality and homotopy theories. Alexander duality states that the Čech cohomology of a compactum X in S^n is canonically isomorphic to the ordinary homology with compact supports of the complement $S^n \setminus X$. In [St-1] Steenrod defined a homology theory ${}^{S}H_*$ such that ${}^{S}H_*(X)$ is canonically isomorphic to the ordinary cohomology of $S^n \setminus X$. ${}^{S}H_*$ is now called (ordinary) Steenrod homology. Milnor [Mil-1] (see also [Sky]) showed that ${}^{S}H_*$ satisfies all of the Eilenberg-Steenrod [E-S] axioms on CM, and further that ${}^{S}H_*$ is characterized by these axioms together with two additional axioms; namely, invariance under relative homeomorphism (generalized excision) and the strong wedge axiom (see (8.2.3)).

(8.2.2) Steenrod K-homology. In a series of papers [Brow], [B-D-F-1-2], L. Brown, R. Douglas, and P. Fillmore defined a functor Ext on CM by taking for Ext (X) unitary equivalence classes of C^*-algebra extensions of the compact operators by the C^*-algebra of complex-valued functions on X. Kaminker and Schochet [K-S] then set

$$^{S}E_n(X) \equiv \begin{cases} \text{Ext}(X), & \text{for } n \text{ odd}, \\ \text{Ext}(\Sigma X), & \text{for } n \text{ even}, \end{cases}$$

(Σ is the unreduced suspension) and showed that $^{S}E_*$ satisfies the axioms (8.2.3) for a generalized reduced Steenrod homology theory. L. Brown, Douglas, and Fillmore have shown that on the category of finite complexes, $^{S}E_*$ is reduced K-homology.

(8.2.3) The Kaminker-Schochet [K-S] axioms. A generalized (reduced) Steenrod homology theory consists of a sequence $h_* = \{h_n | n \in Z\}$ of covariant, homotopy invariant functors from the category CM of compact metric spaces to the category AG of abelian groups which satisfy the following axioms.

(E) Exactness: if (X,A) is a compact metric pair, then the natural sequence

$$h_n(A) \longrightarrow h_n(X) \longrightarrow h_n(X/A)$$

is exact for all n.

(S) Suspension: there is a sequence of natural equivalences

$$\sigma_n : h_n \longrightarrow h_{n+1} \circ \Sigma$$

called suspension, where Σ is unreduced suspension.

(W) Strong wedge: if $X = \vee_{n=1}^{\infty} X_j \equiv \lim_N \{ \vee_{n=1}^{N} X_j \}$ is the strong wedge of a sequence of pointed compact metric spaces, then the natural projections $X \to X_j$ induce an isomorphism

$$h_*(X) \cong \Pi_j h_*(X_j).$$

(8.2.4) Remarks. The homology theory h_* is reduced (see axiom (E) above) but not pointed. A reduced theory $\tilde{k}_*$ may be obtained from any unreduced theory k_* by the formula $\tilde{k}_n(X) = \ker (k_n(X) \longrightarrow k_n(*))$.

We break the problem of construction Steenrod extensions into two parts. The first part involves approximating a compactum by an inverse system of simplicial complexes, an idea which goes back to Alexandroff [Alex]. The second part involves prolonging a generalized homology theory defined on the category of finite CW complexes to a generalized homology theory defined on pro - SS and satisfying suitable analogues of the Kaminker-Schochet axioms. This was done in §5.6.

We shall carry out the first part of this program using a Vietoris functor (see (8.2.7); this construction was first introduced by T. Porter [Por -1]) after giving Steenrod's original construction for motivation. The Vietoris functor also yields a Steenrod homotopy type for any compact metric space.

We can construct Steenrod theories in a wide variety of other settings, for example, algebraic geometry (étale homotopy theory), proper homotopy theory at ∞, and algebraic K-theory, by applying suitable functors into pro-SS and then following the second part of our program.

(8.2.5) Regular cycles following Steenrod [St-1]. Let X be a compact metric space, K an abstract countable locally finite simplicial complex (clf simplicial complex), and let V_K be the set of vertices of K. A regular map is a map $f:V_K \longrightarrow X$ such that for each $\varepsilon > 0$, the images of all but finitely many simplices have diameter less than ε. A regular q-chain on X with coefficients in an abelian group G, (K,f,c_q), consists of a clf simplicial complex K and a regular map f as above, together with a (possibly infinite) q-chain c_q on K with coefficients in G. One then obtains a chain complex $C_*^R(X;G)$ based upon regular chains, and reduced Steenrod homology

$$^{S}\tilde{H}_q(X;G) \equiv H_{q+1}\left(C_*^R(X;G)\right).$$

(8.2.6) Remarks. Consider a regular q-chain (K,f,c_q) on a compact metric space X. Let $\mathcal{U}$ be an open cover of X. Then there is an $\varepsilon > 0$ (the Lebesque number of $\mathcal{U}$) such that for any point x in X, the ε-neighborhood of x is contained within a single open set $U \in \mathcal{U}$. Hence, for almost all simplices Δ of K, the image under f of the vertices of K is contained within some open set U (depending upon Δ) $\in \mathcal{U}$. Also, the fact that f maps V_K (and not K itself) to X conceals the "local pathology" of X.

(8.2.7) <u>The Vietoris construction</u> [Por-1]. Let $\mathcal{U}$ be an open covering of a topological space X. The <u>Vietoris nerve</u> of $\mathcal{U}$, denoted $VN(\mathcal{U})$, is the <u>simplicial set</u> in which an n-simplex is an ordered (n+1)-tuple $(x_0,x_1,\cdots,x_n)$ of points contained in an open set $U \in \mathcal{U}$. Faces and degeneracies are given by

$d_i(x_0,x_1,\cdots,x_n) = (x_0,x_1,\cdots,x_{i-1},x_{i+1},\cdots,x_n)$, and

$s_i(x_0,x_1,\cdots,x_n) = (x_0,x_1,\cdots,x_{i-1},x_i,x_i,x_{i+1},\cdots,x_n)$, for $0 \leq i \leq n$.

Now consider open covers $\mathcal{U}$ and $\mathcal{V}$ of X where $\mathcal{V}$ refines $\mathcal{U}$ (<u>notation</u>: $\mathcal{U} \leq \mathcal{V}$). The identity map of X induces a canonical inclusion $VN(\mathcal{V}) \hookrightarrow VN(\mathcal{U})$. We may therefore associate to a topological space its <u>Vietoris complex</u>.

(8.2.2) $VX = \{VN(\mathcal{U}) | \mathcal{U}$ an open cover of $X\}$. VX is bonded by the canonical inclusions $VN(\mathcal{V}) \hookrightarrow VN(\mathcal{U})$ when $\mathcal{U} \leq \mathcal{V}$.

(8.2.9) <u>Proposition</u>. The Vietoris construction extends to a functor $V: Top \longrightarrow pro-SS$.

<u>Proof</u>. We define V on morphisms as follows: to a continuous map $f: X \longrightarrow Y$ and open cover $\mathcal{U}$ of Y we associate the open cover

$f^{-1}(\mathcal{U}) = \{f^{-1}(U) | U \in \mathcal{U}\}$ of X. Then f induces an obvious map

$VN(f^{-1}(\mathcal{U})) \longrightarrow VN(\mathcal{U})$ $((x_0,x_1,\cdots,x_n) \longmapsto (f(x_0),f(x_1),\cdots,f(x_n))$ such that whenever $\mathcal{U} \leq \mathcal{V}$ (and hence $f^{-1}(\mathcal{U}) \leq f^{-1}(\mathcal{V})$), the diagrams

$$\begin{array}{ccc} VN(f^{-1}(\mathcal{V})) & \longrightarrow & VN(\mathcal{V}) \\ \big\downarrow & & \big\downarrow \\ VN(f^{-1}(\mathcal{U})) & \longrightarrow & VN(\mathcal{U}) \end{array}$$

commute. These diagrams induce the required morphism $Vf : VX \longrightarrow VY$ in pro-SS. Clearly V preserves identity maps and $V(fg) = Vf \circ Vg$. □

We shall frequently use the following theorem of Dowker.

(8.2.10) <u>Theorem</u> [Dow]. Let $\mathcal{U}$ be an open covering of a topological space X. Then the realization of the Vietoris nerve of $\mathcal{U}$, $RVN\mathcal{U}$, is canonically homotopy equivalent to the realization of the Čech nerve of $\mathcal{U}$, $RCN\mathcal{U}$. □

We need Vietoris nerves rather than Čech nerves in order to obtain a functor from compact metric spaces to pro-spaces. An interesting problem is the construction of a nerve that is "small" like the Čech nerve and "rigid" like the Vietoris nerve.

(8.2.11) <u>Steenrod and Čech extensions of homology theories</u>. Let h_* be a generalized (reduced) homology theory defined on the category of finite CW complexes. By G. W. Whitehead [Wh] there is a CW spectrum E which represents h_*, i.e., $h_*(-) \cong \pi_*{}^s((-) \wedge E)$. See also §2.2 and (5.6.7). Define the Steenrod and Cech extensions of h_* to the category CM of compact metric spaces by the following formulas. [We write ${}^S h_*$ for both the extension of h_* to pro-SS (5.6.7) and the Steenrod extension; the usage should be clear from the context.)

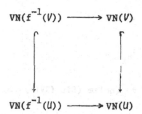

(8.2.12) Steenrod extension:

$$^{S}h_*(-) \equiv {}^{S}h_* \circ V$$

$$= \pi_*^{s}(\mathrm{holim}\ (\mathrm{Sin}\ (RV(-) \wedge E))$$

$$= \mathrm{Ho}(\mathrm{pro} - \mathrm{Sp})(S^*, \mathrm{Sin}\ (RV(-) \wedge E)).$$

(8.2.13) Čech extension:

$$\check{h}_*(-) = \lim_j \{h_* V(-)_j\}$$

$$= \lim_j \{\pi_*^{s}(\mathrm{Sin}\ (RV(-)_j \wedge E)\}$$

$$= \mathrm{pro} - \mathrm{Ho}(\mathrm{Sp})(S^*, \mathrm{Sin}\ (RV(-) \wedge E)).$$

Here, $V(-) = \{V(-)_j\}$ denotes the Vietoris functor and Sp the category of simplicial spectra.

For ordinary homology these formulas become

(8.2.14)
$$^{S}\tilde{H}_*(-;R) \equiv {}^{S}\tilde{H}_*(V(-);R)$$

$$= \pi_*(\mathrm{holim}\ RV(-))$$

$$= \mathrm{Ho}(\mathrm{pro} - SS_*)(S^*, RV(-));$$

(8.2.15)
$$\check{\tilde{H}}_*(-;R) = \lim_j \{\tilde{H}_*(V(-)_j;R)\}$$

$$= \lim_j \{\pi_*(RV(-)_j)\}$$

$$= \mathrm{pro} - \mathrm{Ho}(SS_*)(S^*, RV(-)).$$

In (8.2.14) and (8.2.15), R denotes any commutative ring with identity as well as the free R-module functor of Bousfield and Kan [B-K-1], <u>not</u> the geometric realization functor.

(8.2.16) <u>Remarks</u>. Ken Brown [Brown] defined generalized <u>sheaf cohomology theories</u> with a similar use of simplicial spectra and smash products.

(8.2.17) <u>Theorem</u>. $\check{h}_*$ is the Čech extension of h_*.

<u>Proof</u>. This follows from Dowker's Theorem (see (8.2.10)). Alternatively, follow the proofs of Theorems (8.2.18) and (8.2.21). □

(8.2.18) <u>Theorem</u>. ${}^{S}h_*$ is a homology theory on the category CM of compact metric spaces.

(8.2.19) <u>Theorem</u>. ${}^{S}h_* \cong h_*$ on finite CW complexes.

(8.2.20) <u>Theorem</u>. Products and operations associated with h_* extend to ${}^{S}h_*$.

(8.2.21) <u>Theorem</u>. ${}^{S}h_*$ is a Steenrod homology theory on the category of compact metric spaces.

(8.2.22) <u>Theorem</u> (see also [K-K-S]). ${}^{S}h_*$ satisfies functional and Steenrod duality on compact metric spaces.

We begin the proofs by verifying properties of the Vietoris functor in §8.4. We shall prove Theorem (8.2.18) in §8.4 using strong homology theories on pro-SS,

(5.6.7). Theorems (8.2.19) and (8.2.20) follow easily, see §8.3. We shall develop a Bousfield-Kan spectral sequence for ${}^{S}h_*$ in §8.5 (compare [Brown]) and prove Theorem (8.2.21) there. We shall prove Theorem (8.2.22) in §8.6.

§8.3. The Vietoris functor

We shall prove that the Vietoris functor on the category CM of compact metric spaces preserves homotopies, cofibration sequences, suspensions, and limits, at least up to canonical equivalence in Ho(pro - SS).

(8.3.1) Proposition (announced independently by T. Porter [Por - 1]). Homotopic maps of spaces $f,g:X \rightrightarrows Y$ induce homotopic maps in pro - SS, $Vf,Vg:VX \rightrightarrows VY$, of their Vietoris systems, hence V induces a functor

$$V:\mathrm{Ho(Top)} \longrightarrow \mathrm{Ho(pro-SS)}.$$

Proof. Let $H:X \times [0,1] \longrightarrow Y$ be a homotopy with $Hi_0 = f$ and $Hi_1 = g$, where i_0 and i_1 are the inclusions of X as the ends of the cylinder $X \times [0,1]$. It suffices to show that $Vi_0 = Vi_1$ in Ho(pro - SS). We shall define a homotopy K from RVi_0 to RVi_1 in pro - Top, i.e., a map

$$K:RVX \times [0,1] \longrightarrow RV(X \times [0,1])$$

with $K_0 = RVi_0$ and $K_1 = RVi_1$ (the realization functor R is applied levelwise). Because adjunction morphisms $id \longrightarrow \mathrm{Sin}\, R$ are natural weak equivalences in SS, we obtain a diagram

$$(8.3.2) \qquad VX \times [0,1] \longrightarrow \text{Sin}\,(RVX \times [0,1])$$

$$\longrightarrow \text{Sin}\,RV(X \times [0,1])$$

$$\xleftarrow{\simeq} V(X \times [0,1]).$$

Because the "wrong-way" arrow in diagram (8.3.2) is invertible in Ho(pro-SS), we shall see that $Vi_0 \simeq Vi_1$, as required.

Call an open covering $\mathcal{U}$ of $X \times [0,1]$ a <u>stacked covering</u> if $\mathcal{U}$ is a union of families of open sets $U \times V_\alpha$ where U is an open set in X and V_α is an open covering of $[0,1]$ depending upon U (see [E-S]).

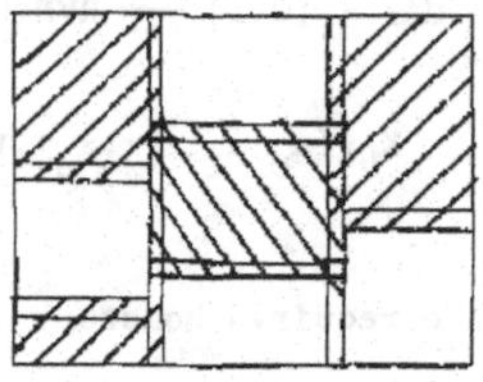

Let $\mathcal{U}$ be a covering of $X \times [0,1]$ by basic open sets, i.e.,

$$\mathcal{U} = \{U_\alpha \times V_\alpha | U_\alpha \text{ open in } X, \quad V_\alpha \text{ open in } [0,1]\}.$$

Such coverings are clearly cofinal in the inverse system of all coverings of $X \times [0,1]$. For each x in X, consider the induced covering $\mathcal{U}_x$ of $x \times [0,1] \subset X \times [0,1]$. Because $[0,1]$ is compact, $\mathcal{U}_x$ admits a finite subcover, say $\{U_{x,i} \times V_{x,i} | i = 1,2,\cdots,n_x\}$. Let $U_x = \bigcap_{i=1}^{n_x} U_{x,i}$, and form the stacked covering

$$\mathcal{U}' = \{U_x \times V_{x,i} | i = 1,2,\cdots,n_x, \quad x \in X\}.$$

Clearly each open set of $\mathcal{U}'$ is contained in an open set of $\mathcal{U}$. Hence, stacked coverings are cofinal in all coverings of $X \times [0,1]$.

To define K, let $\mathcal{U}$ be a stacked covering of $X \times [0,1]$, say

(8.3.3) $$\mathcal{U} = \cup_{U \in \mathcal{U}'}\{U \times V_\alpha | \{V_\alpha\} \text{ is an open covering of } [0,1]\}$$

where $\mathcal{U}'$ is an open covering of X. Then the homotopies

$$K_{\mathcal{U}}: RVN(\mathcal{U}') \times [0,1] \longrightarrow RVN(\mathcal{U}), \quad \text{with}$$

$$K_{\mathcal{U}}((x_0,x_1,\cdots,x_n),t) = ((x_0,t),(x_1,t),\cdots,(x_n,t))$$

from RVi_0 to RVi_1 form the required homotopy

$$K = \{K_{\mathcal{U}}: RVN(\mathcal{U}') \times [0,1] \longrightarrow RVN(\mathcal{U}) | \mathcal{U} \text{ a stacked covering of } X, \ \mathcal{U}' \text{ as in (8.3.3)}\}$$

in pro-Top with $K_0 = RVi_0: RVX \longrightarrow RV(X \times [0,1])$ and $K_1 = RVi_1: RVX \longrightarrow RV(X \times [0,1])$. □

(8.3.4) Proposition. Let A be a closed subset of a topological space X. Then the induced map $VA \longrightarrow VX$ is a cofibration in pro-SS.

Proof. We may represent the map $VA \longrightarrow VX$ as a levelwise cofibration in an appropriate level category SS^J as follows. Each open covering $\mathcal{U}$ of X induces an open covering $\mathcal{U}|A$ of A, namely

(8.3.5) $$\mathcal{U}|A = \{U \cap A \mid U \in \mathcal{U}\},$$

and an inclusion of Vietoris nerves

$$VN(\mathcal{U}|A) \longrightarrow VN(\mathcal{U}).$$

Because each open covering of A can be extended to an open covering of X by adjoining the open set $X \setminus A$, the set of restrictions of open coverings of X to A, $\{\mathcal{U}|A\}$ (see (8.3.5)), is cofinal in the set of all open coverings of A. We obtain the required representation

$$VA \cong \{VN(\mathcal{U}|A) \mid \mathcal{U} \text{ an open covering of } X\} \hookrightarrow \{VN\} \equiv VX. \quad \square$$

(8.3.6) Proposition. Given $A \subset X$, there is a natural map $VX/VA \longrightarrow V(X/A)$.

Proof. In the solid-arrow diagram

$$\begin{array}{ccccc} VA & \longrightarrow & VX & \longrightarrow & VX/VA \\ & & & \searrow & \vdots \\ & & & & V(X/A), \end{array}$$

the composite mapping $VA \to VX \to V(X/A)$ is trivial. This yields the required map. $\square$

(8.3.7) Proposition. Let A be a closed subset of a compact metric space X. Then there is a natural equivalence $VX/VA \longrightarrow V(X/A)$ in Ho(pro − SS), hence the sequence $VA \to VX \to V(X/A)$ is a cofibration sequence in Ho(pro − SS).

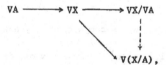

The following lemma about "shape cofibrations" is a key tool in the proof. It is analogous to the statement that a map $A \longrightarrow X$ is a cofibration if and only if there is a neighborhood N of A in X such that A is a strong deformation retract of N [St - 3]. We state it in somewhat greater generality than is needed now; we use the extra generality in Proposition (8.3.22), below.

(8.3.8) Lemma. Let A be a closed subset of a compact metric space X, and let $\mathcal{U}$ be an open covering of X. Then there is an open covering $\mathcal{V}$ of X and a neighborhood N of A in X with the following properties:

a) $\mathcal{V}$ refines $\mathcal{U}$;

b) For each open set V of $\mathcal{V}$, either $V \cap A = \phi$ or $V \subset N$;

c) For each neighborhood N' of A in N, the inclusion of Vietoris nerves

$$VN(\mathcal{V}|A) \longrightarrow VN(\mathcal{V}|N')$$

is an equivalence in $Ho(SS)$.

Proof of Lemma. Because X is compact, we may assume that $\mathcal{U}$ is a finite open cover, $\mathcal{U} = \{U_i | i = 1,2,\cdots,n\}$. From now on, in all constructions and sets indexed by i, i ranges over $\{1,2,\cdots,n\}$. Consider the restriction $\mathcal{U}|A \equiv \{U_i \cap A\}$. By a result of Kuratowski [Kur, p. 122], there are open sets U'_i in X such that $U_i \cap A = U'_i \cap A$, and if $\mathcal{U}' \equiv \{U'_i\}$, the inclusions $U_i \cap A \hookrightarrow U'_i$ induce an isomorphism on Čech nerves

$$CN(\mathcal{U}|A) \xrightarrow{\cong} CN(\mathcal{U}').$$

To do this, let

$$U'_i = \{x \mid d(x, U_i \cap A) < d(x, A \setminus U_i)\}, \tag{8.3.9}$$

Next, perform the following constructions. Let $U''_i = U_i \cap U'_i$, let $U'' = \{U''_i\}$, let $N'' = \cup U''_i$, and, using normality of X, let N be an open set with $A \subset N \subset \overline{N} \subset B''$. Next, let

(8.3.10)

$$V_i^{\,1} = U''_i \cap N,$$

$$V_i^{\,2} = U''_i \cap (N'' \quad A), \text{ and}$$

$$V_i^{\,3} = U_i \cap (X \setminus \overline{N}),$$

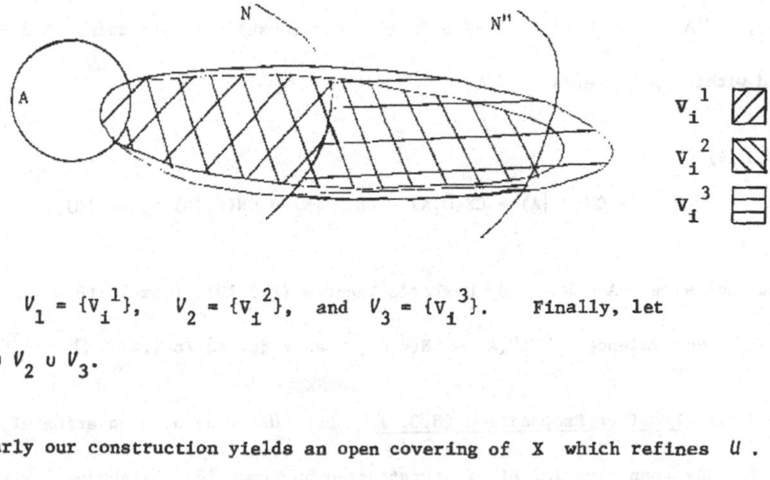

and let $V_1 = \{V_i^{\,1}\}$, $V_2 = \{V_i^{\,2}\}$, and $V_3 = \{V_i^{\,3}\}$. Finally, let $V = V_1 \cup V_2 \cup V_3$.

Clearly our construction yields an open covering of X which refines U. Because the open sets $V_i^{\,1}$ are subsets of N, and the open sets $V_i^{\,2}$ and $V_i^{\,3}$ are disjoint from A by construction, property (b) holds.

To check property (c), first note that the open sets V_i^3 are disjoint from N, hence for any open set N' with $A \subset N' \subset N$,

$$(8.3.11) \qquad V|N' = (V_1 \cup V_2)|N'.$$

Next, observe that each open set $V_i^2 \cap N'$ of $V_2|N'$ is contained in an open set of $V_1|N'$, namely, $V_i^1 \cap N'$. Hence,

$$(8.3.12) \qquad VN(V_1|N') = VN((V_1 \cup V_2)|N') = VN(V|N'),$$

and similarly,

$$(8.3.13) \qquad VN(V_1|A) = VN(V_1 \cup V_2)|A) = VN(V|A).$$

Finally, $V|A = V_1|A = V|A$, and a look at Kuratowski's construction (8.3.9) applied within N' yields

(8.3.14)
$$CN(V|A) = CN(V_1|A) = CN(U|A) \simeq CN(U'|N) \simeq CN(V_1|N) = CN(V|N),$$

via the inclusion $A \subset N$. By Dowker's Theorem (8.2.10), formula (8.3.14) yields the desired equivalence $VN(V|A) \simeq VN(V|N)$, as required in (c). □

(8.3.15) Proof of Proposition (8.3.7). Let U be an open covering of X and V any open covering of X constructed by Lemma (8.3.8) above. With V_2 and V_3 as above, let V' be the open covering $V_2 \cup V_3 \cup \{N\}$ of X. Because N is a neighborhood of A (in X) and the open sets of V_2 and V_3 are disjoint from A, the projection $\pi: X \longrightarrow X/A$ induces an open covering

$V'' \equiv \pi V'$ of X/A. Because V refines V', we obtain a commutative diagram

(8.3.16)

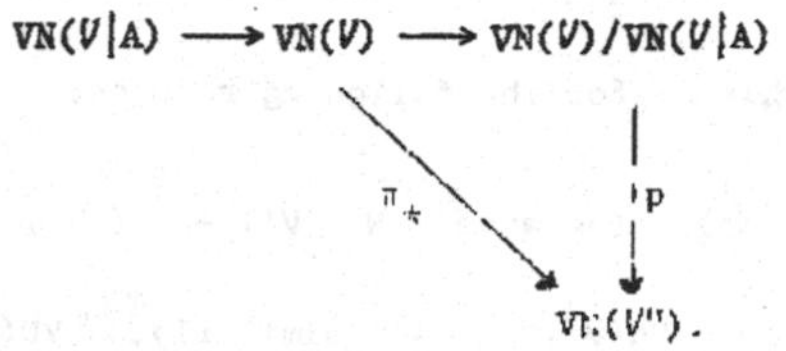

We shall show that p is a weak equivalence.

Form the following commutative diagram.

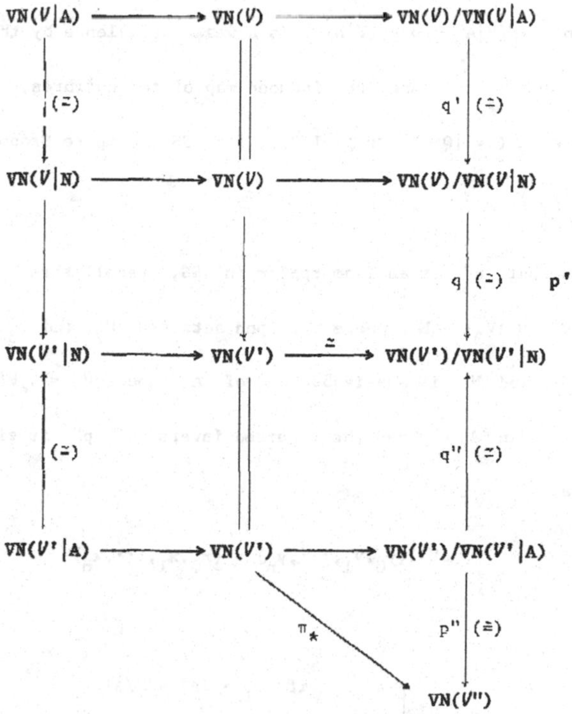

(8.3.17)

In diagram (8.3.17) the composite map $p''p'$ is the map p in diagram (8.3.16), and the rows are cofibration sequences in SS. The indicated maps are weak equivalences (or isomorphisms) for the following reasons.

a) $RVN(V'|N) = RVN(\{N\})$ (because $N \in V'$) $\simeq CN(\{N\})$ (by Dowker's Theorem (8.2.10)) $= *$ so $VN(V'|N) \simeq *$ similarly, $VN(V'|A) \simeq *$. Therefore the maps $VN(V') \longrightarrow VN(V')/VN(V'|N)$ and $VN(V') \longrightarrow VN(V')/VN(V'|A)$ are weak equivalences. Hence q'' is a weak equivalence (by Axiom M5 for SS).

b) The map $VN(V|A) \longrightarrow VN(V|N)$ is a weak equivalence by the construction of V and N. Hence, the induced map of the cofibres, q', is a weak equivalence (by [Q -1, Prop. I.3.5] for SS, compare Proposition (3.4.12)(c)).

c) To show that p'' is an isomorphism in SS, recall that $V' = \{V_2\} \cup \{V_3\}$ N where the open sets of V_2 and V_3 are disjoint from A and N is a neighborhood of A, and $V'' = \pi_* V'$. Choose a point a in A. Then the required inverse of p'' is given by the formula

$$p''^{-1}(y_0,y_1,\cdots,y_n) = (x_0,x_1,\cdots,x_n)$$

where

$$x_i = \begin{cases} a & \text{if } y_i = [A] \in X/A \\ \pi^{-1}y_i & \text{if } y_i \in \pi(X \setminus A) \subset X/A. \end{cases}$$

d) To show that q is a weak equivalence observe that

$$VN(\mathcal{V}') = VN(\mathcal{V}) \cup_{VN(\mathcal{V}|N)} VN(\{N\}) \tag{8.3.18}$$

(and recall that $N \in \mathcal{V}'$ so that $VN(\mathcal{V}'|N) = VN(\{N\})$). Because $VN(\{N\})$ is contractible, the diagram of geometric realizations

$$\begin{array}{ccc} RVN(\mathcal{V}|N) & \longrightarrow & RVN(\{N\}) \\ \Big\downarrow & & \Big\downarrow * \\ RVN(\mathcal{V}|N) & \longrightarrow & CRVN(\mathcal{V}|N) \end{array} \tag{8.3.19}$$

commutes up to homotopy (C denotes the unreduced cone). We may use the homotopy extension property to find a map $RVN(\{N\}) \longrightarrow (RVN(\mathcal{V}|N)$ which makes diagram (8.3.19) strictly commute. This yields a map $RVN(\mathcal{V}') \longrightarrow RVN(\mathcal{V}) \cup_{RVN(\mathcal{V}|N)} (RVN(\mathcal{V}|N)$ (R preserves cofibrations, cones, and quotients by [Mil -2]) which extends the identity map of $RVN(\mathcal{V})$ and is a weak equivalence by an easy argument involving the homotopy extension property. Because $VN(\mathcal{V}') \simeq VN(\mathcal{V}')/VN(\mathcal{V}'|N)$ (see (a), above), the map q of diagram (8.3.17) is a weak equivalence, as required.

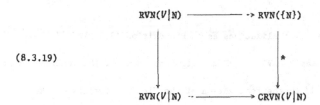

It follows that the map p in diagram (8.3.16) is a weak equivalence. It is easy to check that open coverings of X/A of the form $\mathcal{V}''$ constructed above are cofinal in all open coverings of X/A. Therefore, the map $VX/VA \longrightarrow V(X/A)$ factors as a level weak equivalence (use the maps p of diagrams (8.3.8)) followed by a cofinal inclusion $\{VN(V'')\} \subset V(S/A)$. Thus $VX/VA \simeq V(X/A)$ as required. □

(8.3.20) <u>Proposition</u>. There are natural weak equivalences $\Sigma\, VX \longrightarrow V\, \Sigma\, X$ in Ho(pro-Top), where Σ denotes the appropriate unreduced suspension.

<u>Proof</u>. We use geometric realizations as in Proposition (8.3.1). Following (8.3.1), we may define a map $CRVX \longrightarrow RVCX$ (from the cone of the realization of the Vietoris complex of X to the realization of the Vietoris complex of the cone of X) which yields a commutative solid arrow diagram

(8.3.21)
$$\begin{array}{ccccc} RVX & \dashrightarrow & CRVX & \longrightarrow & CRVX/RVX \\ \| & & \downarrow & & \downarrow \\ RVX & \longrightarrow & RVCX & \dashrightarrow & RVCX/RVX \end{array}$$

in which the rows are cofibration sequences and the vertical maps are equivalences in Ho(pro-Top). Proposition (3.4.12) yields a filler (in Ho(pro-Top)) $CRVX/RVX \longrightarrow RVX/RVCX$ in diagram (8.3.21) which is also an equivalence there. But $CRVX/RVX \equiv \Sigma\, RVX \cong R\, \Sigma\, VX$ (R preserves suspensions by [Mil-2]) and $RVCX/RVX = R(VCX/VX)$ (R preserves quotients by [Mil-2]) $\simeq RV(CX/X)$ (by Proposition (8.3.7) $\equiv RV\, \Sigma\, X$. Naturality is easy to check. The conclusion follows. □

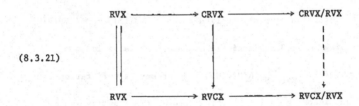

(8.3.22) Proposition. Let $\{X_j\}$ be an inverse system of compact metric spaces, and let $X = \lim_j \{X_j\}$. Then the projections $X \longrightarrow X_j$ induce a natural equivalence $VX \xrightarrow{\simeq} \{VX_j\}$ in Ho(pro - SS).

Proof. By first applying the Mardešic trick (Theorem (2.1.6)) if necessary, we may assume that the indexing category $J = \{j\}$ is a cofinite strongly directed set.

Let $\{\mathcal{U}_{j,k} | k \in K_j\}$ be the inverse system of all finite open coverings of X_j, where the k - index is assigned so that for $j' > j$, $\mathcal{U}_{j',k}$ is the pullback of $\mathcal{U}_{j,k}$ to $X_{j',k}$. Again, by applying the Mardešic trick if necessary, we may assume that each indexing category K_j, where $j \in J$, is a cofinite strongly directed set. Because each X_j is compact, $\{VN(\mathcal{U}_{j,k})\}$ is cofinal in the Vietoris system of X_j, hence isomorphic to it. Assign a partial order to $\cup_{j \in J}\{j\} \times K_j$ as follows: $(j,k) \leq (j',k')$ if $j \leq j'$ and $k \leq k'$. Then

(8.3.23) $$\lim_j\{VX_j\} \cong \lim_j\{\{VN(\mathcal{U}_{j,k})\}_{k \in K_j}\} = \{VN(\mathcal{U}_{j,k})\}_{j \in J,\ k \in K_j}$$

(an inverse system is its own limit in any pro - category). From now on, unless otherwise stated, (j,k) ranges over $\cup_{j \in J}\{j\} \times K_j$.

We shall now write the natural map $VX \longrightarrow \lim_j VX_j$ as a composite of several maps which will be later shown to be weak equivalences. To do this, let X'_j denote the image of X in X_j. Apply Lemma (8.3.8) inductively to obtain open coverings $\mathcal{V}_{j,k}$ of X_j and neighborhoods $N_{j,k}$ of X'_j in X_j with the following properties:

a) $\mathcal{V}_{j,k}$ refines $\mathcal{V}_{j,k'}$ for $k' < k$, yielding inverse systems $\{\mathcal{V}_{j,k}\}_{k\in K_j}$ for $j \in J$;

b) $\mathcal{V}_{j,k}$ refines $\mathcal{U}_{j,k}$, so that $\{\mathcal{V}_{j,k}\}_{k\in K_j}$ is cofinal in $\{\mathcal{U}_{j,k}\}_{k\in K_j}$ for $j \in J$;

c) $\mathcal{V}_{j,k}$ refines $\mathcal{V}_{j',k}$ for $j' < j$, yielding an inverse system $\{\mathcal{V}_{j,k}\}$ which is cofinal in $\{\mathcal{U}_{j,k}\}$ by (b);

d) $N_{j,k} \subset N_{j,k'}$ for $k' < k$, yielding inverse systems $\{VN(\mathcal{V}_{j,k}|N_{j,k})\}_{k\in K_j}$ for $j \in J$;

e) $N_{j,k}$ is contained in the pullback of $N_{j',k}$ to $X_{j,k}$ for $j' < j$, yielding an inverse system $\{VN(\mathcal{V}_{j,k}|N_{j,k})\}$;

f) The inclusions $VN(\mathcal{V}_{j,k}|X'_j) \hookrightarrow VN(\mathcal{V}_{j,k}|N_{j,k})$ are equivalences in $Ho(SS)$, hence the levelwise inclusion $\{VN(\mathcal{V}_{j,k}|X'_j)\} \hookrightarrow \{VN(\mathcal{V}_{j,k}|N_{j,k})\}$ is an equivalence in $Ho(pro-SS)$.

Factor the natural map $VX \longrightarrow \lim_j \{VX_j\}$ as follows:

$$(8.3.24) \qquad \begin{aligned} VX &\xrightarrow{\pi} \{VN(\mathcal{U}_{j,k}|\overline{X}_j)\} \\ &\cong \{VN(\mathcal{V}_{j,k}|\overline{X}_j)\} \quad \text{(by (c), above)} \\ &\xrightarrow{\cong} \{VN(\mathcal{V}_{j,k}|N_{j,k})\} \quad \text{(by (f), above)} \\ &\xrightarrow{i} \{VN(\mathcal{V}_{j,k})\} \\ &\cong \lim_j\{\{VN(\mathcal{V}_{j,k})\}_{k\in K_j}\} \\ &\cong \lim_j\{VX_j\} \quad \text{(by (b), above).} \end{aligned}$$

We shall complete the proof by observing that the maps π and i above are pro-isomorphisms. For π this is an easy consequence of compactness and properties of the product topology. For i, consider any fixed index (j,k). The images of $X_{j'}$ in X_j for $j' > j$ form a family of compact sets whose intersection X'_j is contained in the open set $N_{j,k}$. Hence we may choose a $j' > j$ such that the image of $X_{j'}$ in X_j is contained within $N_{j,k}$. Because $\mathcal{V}_{j',k}$ refines $\mathcal{V}_{j,k}$ (property (c), above), we obtain a commutative diagram

$$\begin{array}{ccc} VN(\mathcal{V}_{j',k}|N_{j',k}) & \hookrightarrow & VN(\mathcal{V}_{j',k}) \\ \downarrow & \swarrow & \downarrow \\ VN(\mathcal{V}_{j,k}|N_{j,k}) & \hookrightarrow & VN(\mathcal{V}_{j,k}) \end{array}.$$

Hence, i is a pro-isomorphism, as required. The conclusion follows. □

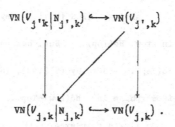

§8.4. Proofs of Theorems (8.2.19), (8.2.20), and (8.2.21).

(8.4.1) Proof of Theorem (8.2.19). Because V preserves homotopies (Proposition (8.3.1)) and the strong homology theory ${}^{S}h_{*}$ on pro-SS is homotopy invariant (5.6.7), the composites ${}^{S}h_{n} \circ V$ are homotopy invariant. Similarly, because V preserves cofibration sequences (Proposition (8.3.7)), exactness of ${}^{S}h_{n} \circ V$ follows from exactness of ${}^{S}h_{n}$ (5.6.7). Thus, Axiom (8.2.3)(E) holds. For Axiom (S), the required natural equivalences are given by

$${}^{S}h_{n} \circ V \xrightarrow{\cong} {}^{S}h_{n+1} \circ \Sigma \circ V \quad \text{(by (5.6.7))}$$

$$\xrightarrow{\cong} {}^{S}h_{n+1} \circ V \circ \Sigma \quad \text{(by (5.6.7) and Proposition (8.3.20)).}$$

The conclusion follows. □

(8.4.2) Proof of Theorem (8.2.20). Let X be a finite complex. Then VX admits a cofinal subtower $\{VN(\mathcal{U}_n)\}$ where each $\mathcal{U}_n$ is a finite open covering of X. Let $\{RCN\{\mathcal{U}_n\}\}$ be the tower of Čech nerves. Choose bonding maps $RCN\{\mathcal{U}_{n+1}\} \longrightarrow RCN\{\mathcal{U}_n\}$ to rigidify $\{RCN\{\mathcal{U}_n\}\}$. By [Dow], (see (8.2.10), $\{RVN\{\mathcal{U}_n\}\} = \{RCN\{\mathcal{U}_n\}\}$ in tow-Ho(Top). But, because X is a complex, $\{RCN\{\mathcal{U}_n\}\} \cong X$ in tow-Ho(Top). But, by (5.2.13), the composite mapping $\{RVN\{\mathcal{U}_n\}\} \longrightarrow X$ is an isomorphism in Ho(tow-Top). This yields a natural isomorphism $\eta: {}^{S}h_{*} \longrightarrow h_{*}$ on finite complexes. □

(8.4.2) Proof of Theorem (8.2.21). Interpret products and operations in terms of maps of spectra (see [Adams-1] or §2.2). The conclusion follows easily from our formula. □

§8.5. Spectral sequences.

We develop Bousfield-Kan and Atiyah-Hirzebruch type spectral sequences which converge to ${}^{S}h_*$. Ken Brown [Brown] developed similar spectral sequences for sheaf cohomology; Kaminker and Schochet [K - S, Theorem (3.10)] obtained the second spectral sequence using fundamental complexes. We shall use the Bousfield-Kan spectral sequence to verify that ${}^{S}h_*$ satisfies Axiom (8.2.3(W)). This completes the proof that ${}^{S}h_*$ is a generalized (reduced) Steenrod homology theory.

(8.5.1) Theorem. (Bousfield-Kan spectral sequence). Let $\{X_j\}$ be an inverse system of compact metric spaces and let $X = \lim \{X_j\}$. Then there is a spectral sequence with

$$E_2^{p,q} = \lim^p{}_j\{{}^{S}h_q(X_j)\},$$

which converges completely under suitable circumstances to ${}^{S}h_*(X)$.

Proof. Recall that $VX \cong \{VX_j\}$ in pro - SS. Now apply the Bousfield-Kan spectral sequence (4.9.4) for ${}^{S}h_q$ applied to the inverse system $\{VX_j\}$ in "pro - (pro - SS)". Compare Proposition (5.6.8). The conclusion follows. □

In particular, suppose that X_j is an inverse system of cardinality $\leq \aleph_n$ so that $\lim^p{}_j\{{}^{S}h_q(X_j)\} = 0$ unless $0 \leq p \leq n+1$. Then $E_{n+2}^{p,q} = E_\infty^{p,q}$ because for $r > n+1$ the differentials d_r (of bidegree $(r, r-1)$) either begin or end at a 0 -group. This is complete convergence. We cite an important special case.

(8.5.2) Corollary. (Compare (4.9.3)). Let $\{X_j\}$ be a tower of compact metric spaces and $X = \lim \{X_j\}$. Then there are short exact sequences.

$$0 \longrightarrow \lim{}^1_j \{{}^Sh_{q+1}(X_j)\} \longrightarrow {}^Sh_q(X) \longrightarrow \lim_j \{{}^Sh_q(X_j)\} \longrightarrow 0. \quad \square$$

(8.5.3) Corollary. The strong wedge axiom ((8.3.3)(W)) holds for ${}^Sh_*$.

Proof. For compact metric spaces $X_1, X_2, \cdots$, consider the tower $\{\bigvee_{j=1}^{N} X_j \mid N = 1,2,\cdots\}$ with $\lim_N \{\bigvee_{j=1}^{N} X_j\} = \bigvee_{j=1}^{\infty} X_j$, the strong wedge. Apply Corollary (8.5.2) to this tower. Because bonding maps ${}^Sh_{q+1}(\bigvee_{j=1}^{N+1} X_j) \longrightarrow {}^Sh_q(\bigvee_{j=1}^{N} X_j)$ in the towers $\{{}^Sh_{q+1}(\bigvee_{j=1}^{N} X_j)\}$ are clearly surjections, the $\lim^1$ terms vanish in this case. The conclusion follows. $\square$

(8.5.4) Remarks.

a) This completes the proof of Theorem (8.2.21).

b) If a compact metric space X is represented as the limit of a tower of polyhedra $\{X_j\}$, Corollary (8.5.2) yields short exact sequences

$$0 \longrightarrow \lim{}^1_j \{h_{q+1}(X_j)\} \longrightarrow {}^Sh_q(X) \longrightarrow \lim_j \{h_q(X_j)\} \longrightarrow 0$$

relating ${}^Sh_*$ to h_*. Compare Milnor's characterization of ordinary (reduced) Steenrod homology [Mil -1]. Uniqueness does not follow in our case because of possible extension problems; however, any natural transformation of Steenrod extensions of h_* is an isomorphism by the above short exact sequences.

(8.5.5) Theorem. (Atiyah-Hirzebruch spectral sequences) [K - S]. Let X be a compact metric space of dimension $d < \infty$. Then there is a spectral sequence with

$$E^2_{p,q} = {}^S\tilde{H}_p(X; h_q(S^0))$$

and d^r of bidegree $(-r, r-1)$ which converges to ${}^Sh_*(X)$ in the sense that $E^{d+1} = E^\infty$.

Our proof is contained in (8.5.6) - (8.5.21), below. For a compact metric space X of dimension $d < \infty$ there is a cofinal tower $\{\mathcal{U}_n\}$ in the Vietoris system VX such that each $\mathcal{U}_n$ is a finite open cover (use compactness) with $\dim CN(\mathcal{U}_n) \leq d$ (use the definition of covering dimension). We therefore begin by proving the following.

(8.5.6) Theorem. Let $X = \{X_n\}$ be a tower of finite simplicial complexes and simplicial maps. Then there is a spectral sequence with

$$E^2_{p,q} = {}^SH_p(X; h_q(S^0))$$

which converges to ${}^Sh_*(X)$ if $\dim X$ is finite. More precisely, if $\dim X \leq d < \infty$, $E^{d+1} = E^\infty$.

Proof. The proof is broken up into several steps: (8.5.7) - (8.5.20), below.

(8.5.7) <u>Construction of the spectral sequence</u>. (Compare [K - S]). Let $F_pX = \{F_pX_n\}$ be the levelwise p-skeleton of X. Following Massey, we define an exact couple

$$\begin{array}{ccc} D^1_{**} & \xrightarrow{\ i\ } & D^1_{**} \\ & {}_{k}\nwarrow \quad \swarrow_{j} & \\ & E^1_{**} & \end{array}$$

where

$$E^1_{p,q} = {}^S h_{p+q}(F_pX/F_{p-1}X),$$

$$D^1_{p,q} = {}^S h_{p+q}(F_pX), \quad \text{and}$$

degree $i = (1,-1)$, degree $j = (0,0)$, and degree $k = (-1,0)$. This yields a spectral sequence $\{E^r_{p,q}(X)\}$.

(8.5.9) <u>Description of E^1 and E^2</u>. For each p and n,

$$(F_pX/F_{p-1}X)_n = (F_pX_n/F_{p-1}X) = \vee S^p,$$

a finite wedge of p-spheres, one for each p-cell of X_n. Hence,

$$E^1_{p,q} = {}^S h_{p+q}\{(\vee S^p)_n\}.$$

Also, because the bonding maps $X_n \longrightarrow X_{n-1}$ are simplicial, the composite maps

$$S^p \xrightarrow{\ i\ } (\vee S^p)_{n+1} \longrightarrow (\vee S^p)_n \xrightarrow{\ \pi\ } S^p$$

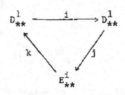

(where i is a typical injection and π a typical projection) have degree 0 except for at most one π for each i; in which case the degree is ± 1. Hence, if ${}^{S}h_*(S^0)$ is a graded ring, we may choose bases for the free modules ${}^{S}h_{p+*}(\vee S^p)_n \cong \bigoplus_{\text{finite}} ({}^{S}h_*(S^0))$ so that the maps

(8.5.10) $\quad {}^{S}h_{p+*}((\vee S^p)_{n+1}) \longrightarrow {}^{S}h_{p+*}((\vee S^p)_n)$ are represented by matrices of

the form

(8.5.11)

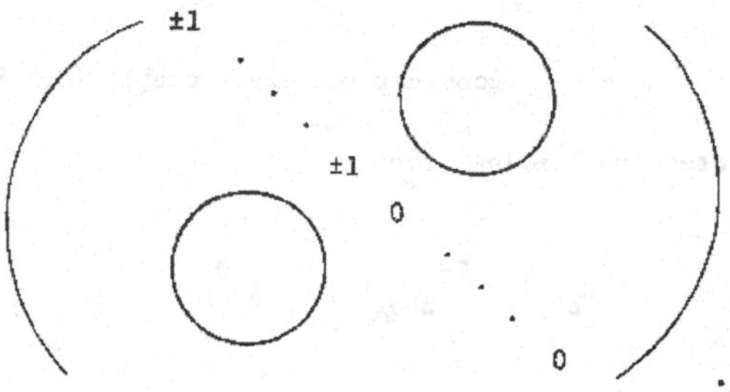

Hence, the towers

(8.5.12) $$\{h_{p+q+1}(\vee S^p)_n \mid n \geq 0\}$$

are Mittag-Leffler, so that

(8.5.13) $$\lim{}^1\{h_{p+q+1}(\vee S^p)_n\} = 0$$

for all p,q. (If ${}^{S}h_*(S^0)$ is not a ring, analogues of (8.5.10) - (8.5.11) still hold, but are more difficult to describe. Thus, (8.5.13) holds in general.) (8.5.13) yields the following useful calculation.

$$(8.5.14)\qquad {}^{S}h_{p+q}\{(\vee S^{p})_{n}\} = h_{p+q}\{(\vee S^{p})_{n}\}$$

$$= \lim_{n}\{h_{p+q}(\vee S^{p})_{n}\}$$

$$= \lim_{n}(\{H_{p}(\vee S^{p})_{n}\};\ h_{q}(S^{0}))$$

$$= \check{H}_{p}(\{(\vee S^{p})_{n}\};\ h_{q}(S^{0}))$$

$$= {}^{S}H_{p}(\{(\vee S^{p})_{n}\};\ h_{q}(S^{0}))$$

Now, for each fixed $q = q_0$ consider the exact couple (8.5.8) associated with the generalized Steenrod homology theory

$$(8.5.15)\qquad {}^{S}k_{n}(-) = {}^{S}\tilde{H}_{n-q_0}(-;\ h_{q_0}(S^{0}))$$

$$\left({}^{S}k_{n}(S^{0}) = \begin{cases} h_{q}(S^{0}) & n = q_0, \\ 0 & n \neq q_0. \end{cases}\right)$$

In this case, the resulting $E^{i}_{p,q} = 0$ unless $q = q_0$, so that in the resulting spectral sequence

$$E^{2}_{p,q}(X) = E^{\infty}_{p,q}(X) = \begin{cases} {}^{S}k_{p+q}(X) = {}^{S}\tilde{H}_{p}(X;\ h_{q}(S^{0})), & \\ 0 & \text{otherwise.} \end{cases}$$

This shows that

(8.5.16) $$E^2_{p,q}(X) \cong {}^S H_p(X; h_q(S^0)),$$

in the original spectral sequence, as required.

(8.5.17) Convergence. If $\dim X = d < \infty$, then $E^1_{p,q} = 0$ unless $0 \leq q \leq d$, so that $d_{d+2} = d_{d+3} = \cdots = 0$ and $E^{d+1}_{p,q} = E^{\infty}_{p,q}$.

(8.5.18) Naturality. Consider a weak equivalence $X \longrightarrow Y$ of towers of simplicial sets of bounded dimension. Then there is a diagram

(8.5.19) $$X \xrightarrow{\simeq} Z \xleftarrow{\simeq} Y,$$

in Tow-SS, hence an isomorphism

(8.5.20)
$$\begin{aligned} E^2_{p,q}(X) &\cong {}^S\tilde{H}_p(X; h_q(S^0)) \\ &\xrightarrow{\cong} {}^S\tilde{H}_p(Z; h_q(S^0)) \\ &\xleftarrow{\cong} {}^S\tilde{H}_p(Y; h_q(S^0)) \\ &\cong E^2_{p,q}(Y). \end{aligned}$$

It is easy to see that the isomorphism (8.5.20) is independent of Z and the maps in (8.5.19). Hence naturality follows from naturality of ${}^S\tilde{H}_*(\,; h_*(S^0))$. This concludes the proof of Theorem (8.5.6). □

(8.5.21) Proof of Theorem (8.5.5). With X a compact metric space of dimension d, choose a cofinal tower of open coverings of X, $\{\mathcal{U}_n\}$, in $V(X)$ such that each $\mathcal{U}_n$ is finite and satisfies $\dim CN(\mathcal{U}_n) \leq d$. Next, choose bonding maps to rigidify the Cech system $CX \equiv \{CN(\mathcal{U}_n)\}$. By Dowker [Dow], see (8.2.10), $RVX \approx RCX$ in tow-Ho(Top). Hence, by (5.2.9), $RVX \approx RCX$ in Ho(tow-Top). Applying Theorem (8.5.6) to $RCX = \{RCN(\mathcal{U}_n)\}$ with $X = \lim_n \{RCN(\mathcal{U}_n)\}$ yields the required spectral sequence. Naturality follows from naturality of the Vietoris construction; replace RCX by RVX in the formula for $E^2_{*,*}(X)$. □

§8.6. Duality.

We shall prove a Steenrod Duality Theorem (8.6.1) for our generalized Steenrod homology theories and a more general theorem (8.6.2) involving functional duality. See §2.2, (2.2.36) - (2.2.43) for Alexander and Spanier-Whitehead duality and functional duality. Kahn, Kaminker and Schochet [K-K-S] took the formula in Theorem (8.6.2) as their definition of Steenrod homology; hence the two definitions agree up to (non-canonical) isomorphism. Our present proof of (8.6.2) and its application to (8.6.1) was motivated by a letter from Kaminker.

(8.6.1) Theorem. Let X be a compactum in S^n. Then

$$ {}^{S}h_p(X) \cong h^{n-p-1}(S^n \setminus X), $$

and the isomorphism is natural with respect to inclusion maps.

We shall first prove the following.

(8.6.2) Theorem. Let X be a compact metric space, and let DX be the functional dual of X. Then

$$ {}^{S}h_p(X) \cong h^{-p}(DX). $$

Theorem (8.6.1) will follow from (8.6.2).

Proof. Let E be the spectrum which represents h_*. Write X as an inverse limit of finite complexes, $X = \lim \{X_j\}$. Let $DX_j \equiv HOM(X_j, S^0)$ in CWSp be the functional dual of X_j. Consider the evaluation maps $DX_j \wedge X_j \longrightarrow S^0$. These induce maps

$$DX_j \wedge X_j \wedge E \longrightarrow S^0 \wedge E,$$

hence by adjointness

$$X_j \wedge E \longrightarrow E^{DX_j},$$

which are weak equivalences because $X_j \simeq DDX_j$ for finite complexes. Thus

$$\begin{aligned} {}^{S}h_p(X) &= \pi_p{}^{S}\{X_j \wedge E\} \\ &\cong \pi_p{}^{S}\{E^{DX_j}\} \end{aligned}$$

We therefore need to show that the natural map

$$\text{hocolim}\,\{DX_j\} \longrightarrow DX$$

(see §4.10 for homotopy colimits) is a weak equivalence. First, consider a class in $\pi_*{}^{s}DX$ represented by a stable map $S^n \longrightarrow HOM(X, S^0)$, or by adjointness, a map $X \wedge S^n \longrightarrow S^0$ or $X \longrightarrow HOM(S^n, S^0) \simeq S^{-n}$. Because each map from X to a polyhedron factors through a nerve (up to homotopy), the map $\pi_*{}^{s}(\text{hocolim}\,\{DX_j\}) \longrightarrow \pi_*{}^{s}DX$ is an epimorphism. Similarly, because each homotopy $X \times I \longrightarrow Y$, Y a polyhedron, factors through a nerve of X, $\pi_*{}^{s}(\text{hocolim}\,\{DX_j\}) \longrightarrow \pi_*{}^{s}DX$ is a monomorphism. Therefore the natural map

$$\operatorname{hocolim}\{DX_j\} \longrightarrow DX$$

is a weak equivalence, so

$$\pi_p{}^s\{E^{DX_j}\} \cong \pi_p{}^s \operatorname{holim}\{E^{DX_j}\}$$

$$\cong \pi_p{}^s E^{\operatorname{hocolim}\{DX_j\}}$$

(E is stable)

$$\cong \pi_p{}^s E^{DX}$$

$$= h^{-p}(DX),$$

as required. □

(8.6.3) Proof of Theorem (8.6.1). If X were a polyhedron, then the $(n-1)$-dual $S^n \setminus X$ would be stably equivalent to the $(n-1)^{st}$ suspension of the functional dual DX. Theorem (8.6.1) would then follow. Instead we choose a sequence $\{X_j\}$ of polyhedral neighborhoods of X in S^n whose intersection is X. The direct systems $\{S^n \setminus X_j\}$ and $\{\Sigma^{n-1}DX_j\}$ are then equivalent in inj-Ho(Sp), hence in Ho(inj-Sp) by the comparison theorem for direct towers ("dual" to (5.2.9)). $h^{n-p-1}(\{S^n \setminus X_j\}) \cong h^{n-p-1}\{\Sigma^{n-1}DX_j\} = h^{-p}\{DX_j\}$ as required. Further, the required constructions in Theorem (5.2.9) and its "dual" can be carried out for pairs, so the isomorphism ${}^S h_p(X) \cong h^{n-p-1}(S^n \setminus X)$ is natural with respect to inclusion maps. □

§9. SOME OPEN QUESTIONS

We shall give a brief list of some open questions.

(9.1) For proving Whitehead Theorems and defining general profinite completions, it would be useful to have a tractible description of coherent prohomotopy theory. R. Vogt [Vogt - 1] gave a geometric description of a category of coherently-homotopy-commutative diagrams and coherent-homotopy classes of morphisms between such diagrams. In [Por - 3], T. Porter has defined a coherent prohomotopy theory copro- Top, and shown that the natural functor from Ho(pro - Top) to copro - Top is a natural equivalence.

(9.2) Off towers, the relationship between Ho(pro - Top) and pro - Ho(Top) is still mysterious. In particular, does every object of pro - Ho(Top) come from an object of pro - Top?

(9.3) If $f:X \longrightarrow Y$ is a morphism of Ho(pro - Top) such that $\pi(f) \in$ pro - Ho(Top) is invertible, then is f invertible? This problem is probably very delicate.

(9.4) Now that the homotopy theory of pro - SS is well developed, one should systematically study completions as pro - objects and compare the theory thus obtained to the usual one (e.g., [B - K]).

(9.5) We expect applications of the machinery of these notes to the problem of classifying open principal G - fibrations (G a compact topological group) (see [Coh]) and to the problem of defining the <u>continuous</u> algebraic K - groups of a topological ring (see [Wag]).

REFERENCES

[Adams-1] J. F. Adams, Stable homotopy and generalized cohomology, Notes by R. Ming, U. Chicago Math Lecture Notes, 1973.

[Adams-2] __________, On the Groups J(X) - IV, Topology 5, (1966), 21-71.

[Adams-3] __________, Lectures on generalized cohomolgoy, in Category theory, homology, and their applications III, Lecture Notes in Math. 99, Springer, Berlin-Heidelberg-New York, 1969, 1-138.

[Alex] P. Alexandroff, Untersuchungen Über Gestalt und Lage Abgeschlossener Mengen, Ann. of Math. 30 (1929), 101-187.

[An-1] D. W. Anderson, Chain functors and homology theories, in Symposium on Algebraic Topology, Lecture Notes in Math. 249, Springer, Berlin-Heidelberg-New York, 1971, 1-12.

[An-2] __________, Simplicial K-theory and generalized homology theories, (mimeographed notes).

[And] R. D. Anderson, Homeomorphisms on Infinite-Dimensional Manifolds, Actes, Congres Intern. Math., (1970), Tome 2, 13-18.

[A-B] __________ and R. Bing, A Complete Elementary Proof that Hilbert Space is Homeomorphic to the Countable Infinite Product of Lines, Bull. Amer. Math. Soc., 74 (1968), 771-792.

[A-M] M. Artin and B. Mazur, Étale Homotopy, Lecture Notes in Math. 100, Springer, Berlin-Heidelberg-New York, 1969.

[Ball] B. J. Ball, Alternative Approaches to Proper Shape Theory, in Studies in Topology, Ed., N. Stravrakas and K. Allen, Academic Press, New York, 1975.

[B-S-1] B. J. Ball and R. Sher, A theory of proper shape for locally compact metric spaces, Bull. Amer. Math. Soc. 79 (1973), 1023-1026.

[B-S-2] ____________________, A theory of proper shape for locally compact metric spaces, Fund. Math. 86 (1974), 163-192.

[B-V] J. M. Boardman and R. M. Vogt, Homotopy-everything H-spaces, Bull. Amer. Math. Soc. 74 (1968), 1117-1122.

[Bor-1] K. Borsuk, Theory of Retracts, Monografie Matematyczne 44, Warszana, 1967.

[Bor-2] ________, Concerning Homotopy Properties of Compacta, Fund. Math. 62 (1968), 223-254.

[Bor-3] ________, Theory of Shape, Monografie Matematyczne 59, Warszawa, 1975.

[B-K] A. K. Bousfield and D. M. Kan, Homotopy Limits, Completions and Localizations, Lecture Notes in Math. 304, Springer, Berlin-Heidelberg-New York, 1973.

[Brown] K. S. Brown, Abstract Homotopy Theory and Generalized Sheaf Cohomology, Trans. Amer. Math. Soc. 186 (1973).

[B-G] __________ and S. Gersten, Algebraic K-Theory as Generalized Sheaf Cohomology, in Lecture Notes in Math. 341, Springer, Berlin-Heidelberg-New York, 1973.

[Br] E. M. Brown, On the Proper Homotopy Type of Simplicial Complexes, in Topology Conference 1973, Ed. R. F. Dickman, Jr., and P. Fletcher, Lecture Notes in Math. 375, Springer, Berlin-Heidelberg-New York, 1974.

[Bro] E. H. Brown, Cohomology Theories, Ann. of Math., 75 (1962), 467-484.

[Brow] L. Brown, Operator Algebras and Algebraic K-Theory, Bull. Amer. Math. Soc. (to appear).

[B-D-F-1] L. Brown, R. Douglas, and P. Fillmore, Unitary Equivalence Modulo the Compact Operators and Extensions of C^*-Algebras, Proc. Conf. on Operator Theory, Lecture Notes in Math. 345, Springer, Berlin-Heidelberg-New York, 1973, 58-128.

[B-D-F-2] ______________________________, Extensions of C^*-Algebras, Operators with Compact Self-Commutators, and K-Homology, Bull. Amer. Math. Soc. 79 (1973), 973-978.

[Buch] D. Buchsbaum, Satellites and Universal Functors, Ann. of Math. 71 (1960), 199-209.

[C-E] H. Cartan and S. Eilenberg, Homological Algebra, Princeton Univ. Press, Princeton, 1956.

[Chap-1] T. A. Chapman, On Some Applications of Infinite Dimensional Manifolds to the Theory of Shape, Fund. Math. 76 (1972), 181-193.

[Chap-4] ___________, All Hilbert Cube Manifolds are Triangulable, (to appear).

[Chap-5] ___________, Simple Homotopy Theory for ANR's, (to appear).

[Chap-6] ___________, Topological Invariance of Whitehead Torsion, Amer. J. of Math., (to appear).

[C-F] ___________ and S. Ferry, Obstruction to Finiteness in the Proper Category, Univ. of Kentucky mimeo, 1975.

[C-S] ___________ and L. Siebenmann, Finding a Boundary for a Hilbert Cube Manifold, (to appear).

[Chr] D. Christie, Net Homotopy for Compacta, Trans. Amer. Math. Soc. 56 (1944), 275-308.

[Coh] J. Cohen, Inverse limits of principal fibrations, Proc. London Math. Soc. (3) 27 (1973), 178-192.

[Cohen] M. Cohen, A course in simple homotopy theory, Graduate Texts in Math. 10, Springer, Berlin-Heidelberg - New York, 1973.

[Con -Fl] P. E. Conner and E. E. Floyd, The relation of cobordism to K- theories, Lecture Notes in Math. 28, Springer, Berlin-Heidelberg - New York, 1966.

[D - T] A. Dold and R. Thom, Quasifaserungen und Unendiche Symmetrische Produkte, Ann. of Math. 67 (1958), 239-281.

[Dow] C. Dowker, Homology Groups of Relations, Ann. of Math., Vol. 56, No. 1, (1952), 84-95.

[D - K] J. Draper and J. Keesling, An Example Concerning the Whitehead Theorem in Shape Theory, (to appear).

[Dror - 1] E. Dror, Pro - Nilpotent Representation of Homology Types, Proc. Amer. Math. Soc. 38 (1973), 657-660.

[Dror - 2] _______, Acyclic Spaces, Topology, (to appear).

[Dyd - 1] J. Dydak, On a Conjecture of Edwards, (to appear).

[Dyd- 2] _______, Some Remarks Concerning the Whitehead Theorem in Shape Theory, Bull. Polon, Acad. Sci. Ser. Sci. Math. Astrom, Phys., (to appear).

[Ed- 1] D. A. Edwards, Étale Homotopy Theory and Shape, in Algebraic and Geometrical Methods in Topology, Ed. L. McAuley, Lecture Notes in Math. 428, Springer, Berlin-Heidelberg - New York, 1974, 58-107.

[E - G - 1] D. A. Edwards and R. Geoghegan, The Stability Problem in Shape and a Whitehead Theorem in Pro - Homotopy, Trans. Amer. Math. Soc. 214, (1975), 261-277.

[E - G - 2] ______________________, Stability Theorems in Shape and Pro-homotopy, Trans. Amer. Math. Soc. (to appear).

[E - G - 3] ______________________, Shapes of Complexes, Ends of Manifolds, Homotopy Limits and the Wall Obstruction, Ann. of Math. 101, (1975), 521-535.

[E - G - 4] ______________________, Infinite-Dimensional Whitehead and Vietoris Theorems in Shape and Pro - Homotopy, Trans. Amer. Math. Soc., (to appear).

[E - G - 5] ______________________, Compacta Weak Shape Equivalent to ANR's, Fund. Math., (to appear).

[E - H - 1] __________ and H. M. Hastings, On Homotopy Inverse Limits and the Vanishing of Lim^s, (mimeographed notes).

[E - H - 2] ______________________, Čech Theory: Its Past, Present, and Future, (mimeographed notes).

[E - H - 3] ______________________, Every Weak Proper Homotopy Equivalence is Weakly Properly Homotopic to a Proper Homotopy Equivalence, Trans. Amer. Math. Soc. (to appear).

[E - H - 4] ______________________, Counterexamples to Infinite-Dimensional Whitehead Theorems in Pro - Homotopy, (mimeographed notes).

[E - H - 5] ______________________, Topological methods in homological algebra, Proc. Amer. Math. Soc., (to appear).

[E - H - W] D. A. Edwards, H. M. Hastings and J. West, Group Actions on Infinite Dimensional Manifolds, (mimeographed notes).

[E - M] ___________ and P. McAuley, The Shape of a Map, Fund. Math., (to appear).

[Edw] R. Edwards, (to appear) (Announced at the Georgia Topology Conference, 1975).

[E - K] S. Eilenberg and G. Kelly, Closed Categories, in Proc. Conf. Categorical Algebra (La Jolla, Calif., 1965), Springer, Berlin-Heidelberg - New York, 1966, 421-562.

[E - S] ___________ and N. Steenrod, Foundations of Algebraic Topology, Princeton Univ. Press, Princeton, 1952.

[F - T - W] F. T. Farrell, L. R. Taylor, and J. B. Wagoner, The Whitehead Theorem in the proper category, Comp. Math. 27 (1973), 1-23.

[Fox] R. Fox, On Shape, Fund. Math. 74 (1972), 47-71.

[Fr - 1] E. Friedlander, Thesis, M.I.T. (1970).

[Fr - 2] ___________, Fibrations in Étale Homotopy Theory, Publ. Math. I.H.E.S., No. 42 (1972), 5-46.

[Fr - 3] ___________, The Étale Homotopy Theory of a Geometric Fibration, Manuscripta Math. 10 (1973), 209-244.

[Fr - 4] ___________, Computations of K - Theories of Finite Fields, Topology 15 (1976), 87-109.

[Fr - 5] ___________, Exceptional Isogenies and the Classifying Spaces of Simple Lie Groups, (to appear).

[Fr - 6] ___________, $K(\pi,1)$'s in Characteristic $p > 0$, Topology (1973), 9-18.

[Fr-7] E. Friedlander, Unstable K-Theories of the Algebraic Closure of a Finite Field, Comment. Math. Helvetici 50 (1975), 141-154.

[G-Z] P. Gabriel and M. Zisman, Calculus of Fractions and Homotopy Theory, Ergebnisse Der Mathematik, Vol. 35, Springer, Berlin-Heidelberg-New York, 1967.

[Gros-1] J. Grossman, A Homotopy Theory of Pro-spaces, Trans. Amer. Math. Soc., (to appear).

[Gros-2] ________, Homotopy Classes of Maps between Pro-Spaces, Mich. Math. J. 21 (1974), 355-362.

[Gro-1] A. Grothendieck, Technique De Decente Et Theorems D'Existence En Geometrie Algebrique I-IV, Seminar Bourbaki, Exp. 190, 195, 212, 221, 1959-60, 1960-61.

[Gro-2] ________, Sur Quelques Points D'Algebre Homologique, Tohoku Math. J. 9 (1957), 119-221.

[Gug] V. K. A. M. Gugenheim, On supercomplexes, Trans. Amer. Math. Soc. 85 (1957), 35-51.

[Has-1] H. M. Hastings, Homotopy Theory of Pro-Spaces I and II, (mimeographed notes, SUNY-Binghamton), 1974.

[Has-2] ________, Fibrations of Compactly Generated Spaces, Mich. Math. J., 21 (1974), 243-251.

[Has-3] ________, A smash product for spectra, Bull. Amer. Math. Soc. 79 (1973), 946-952.

[Has-4] ________, On function spectra, Proc. Amer. Math. Soc. 44 (1974), 186-188.

[Has-5] ________, Stabilizing tensor products, Proc. Amer. Math. Soc. 49 (1975), 1-7.

[Has-6] H. M. Hastings, A higher dimensional Poincare-Bendixson Theorem, (mimeographed notes, 1976).

[Hel-1] A. Heller, Stable Homotopy Categories, Bull. Amer. Math. Soc. 74 (1968), 28-63.

[Hel-2] ________, Completions in Abstract Homotopy Theory, Trans. Amer. Math. Soc. 147 (1970), 573-602.

[Hend] D. Henderson, Infinite-Dimensional Manifolds are Open Subsets of Hilbert Space, Topology 9 (1970), 25-34.

[Hil-1] P. Hilton, Localization in Group Theory and Homotopy Theory and Related Topics, Lecture Notes in Math. 418, Springer, Berlin-Heidelberg-New York, 1974.

[Hus] D. Husemoller, Fibre Bundles, McGraw-Hill Book Co., New York, 1966.

[Jen] C. Jensen, Les Foncteurs Derives de $\varprojlim$ et leurs Applications en Theorie des Modules, Lecture Notes in Math. 254, Springer (1972).

[K-K-S] D. S. Kahn, J. Kaminker, and C. Schochet, Generalized Homology Theories on Compact Metric Spaces, (to appear).

[K-S] J. Kaminker and C. Schochet, K-Theory and Steenrod Homology: Applications to the Brown-Douglas-Fillmore Theory of Operator Algebras, (to appear).

[Kan-1] D. M. Kan, Semisimplicial Spectra, Ill. J. of Math. 7 (1963), 463-478.

[Kan-2] ________, A Combinatorial Definition of Homotopy Groups, Ann. of Math. 67 (1958), 282-312.

[Kan-3] ________, On the k-cochains of a spectrum. Illinois J. Math. 7 (1963), 479-491.

[Kan-4] ________, On C.S.S. complexes, Amer. J. Math. 79 (1957), 449-476.

[K - W] D. M. Kan and G. W. Whitehead, The Reduced Join of Two Spectra, Topology 3 (1965) Suppl. 2, 239-261.

[Kee - 1] J. Keesling, Shape Theory and Compact Connected Abelian Topological Groups, Trans. Amer. Math. Soc., (to appear).

[Kee - 2] ________, The Čech Homology of Compact Connected Abelian Topological Groups with Applications to Shape Theory, (to appear).

[Kur] C. Kuratowski, Topologie II, Monografie Matematyczne, Warszawa, 1948.

[Lub - 1] S. Lubkin, On a Conjecture of Andre Weil, Amer. J. Math. 89 (1967), 443-548.

[Lub - 2] ________, Theory of Covering Spaces, Trans. Amer. Math. Soc., 104 (1962), 205-238.

[Mal] P. Malraison, Homotopy associative categories, in Algebraic and geometrical methods in topology, Ed. L. McAuley, Lecture Notes in Math. 428, Springer, Berlin-Heidelberg - New York 1974, 108-131.

[Mar - 1] S. Mardešic, On the Whitehead Theorem in Shape Theory, Fund. Math., (to appear).

[Mar - 2] ________, Shapes for Topological Spaces, (to appear).

[Mar - 3] ________, A Survey of the Shape Theory of Compacta, General Topology and its Relations to Modern Analysis and Algebra III, Proc. Third Prague Topological Symp. (1971), Academic Press, N. Y., 1973, 291-300.

[Mas] W. S. Massey, Exact couples in algebraic topology I - V. Ann. of Math. 56 (1952, 363-396, 57 (1953), 248-286.

[May - 1] J. P. May, Simplicial Objects in Algebraic Topology, Van Nostrand, Princeton, N. J., 1967.

[May - 2] ________, E_∞ - spaces, group completions, and permutative categories, in New Developments in Topology, (Proc, Symp. on Algebraic Topology, Oxford, June 1972), London Math. Society Lecture Notes Series11, 1974, 61-93.

[May - 3] ________, The geometry of iterated loop spaces, Lecture Notes in Math. 271, Springer, Berlin-Heidelberg - New York, 1971.

[Mil - 1] J. Milnor, On the Steenrod Homology Theory, Mimeographed Notes, Berkeley, 1961.

[Mil - 2] ________, The Geometric Realization of a Semi-Simplicial Complex, Ann. of Math. 65 (1957), 357-362.

[Mil - 3] ________, On Axiomatic Homology Theory, Pacific J. Math. 12 (1962), 337-341.

[Mit - 1] B. Mitchell, Theory of Categories, Academic Press, New York, 1965.

[Mit - 2] ________, The Cohomological Dimension of a Directed Set, Canad. J. Math. 25 (1973), 233-238.

[Mor - 1] K. Morita, On Shapes of Topological Spaces, Fund. Math. 86 (1975, 251-259.

[Mor - 2] ________, The Hurewicz and the Whitehead Theorems in Shape Theory, Science Reports of the Tokyo Kyoiku Daigaku, Section A, Vol. 12, No. 346 (1974), 246-258.

[Mor - 3] ________, Another Form of the Whitehead Theorem in Shape Theory, Proceedings of the Japan Academy, 51 (1975), 394-398.

[Mos] M. Moszynska, The Whitehead Theorem in the Theory of Shapes, to appear in Fund. Math.

[Osof] B. Osofsky, The Subscript of $\aleph_n$, Projective Dimension, and the Vanishing on $\underleftarrow{\text{Lim}}^{(n)}$, Bull. Amer. Math. Soc. 90 (1974), 8-26.

[Por-1] T. Porter, Čech Homotopy I, J. London Math. Soc. (2), 6 (1973), 429-436.

[Por-2] ________, Stability Results for Topological Space, Math. Z., (to appear).

[Por-3] ________, Coherent Prohomotopy Theory, (to appear).

[Por-4] ________, Abstract homotopy theory in procategories, (to appear).

[Por-5] ________, Coherent prohomotopical algebra, (to appear).

[Puppe] D. Puppe, On the stable homotopy category, Math. Inst. Heidelberg (mimeo), 1972.

[Q-1] D. Quillen, Homotopical Algebra, Lecture Notes in Math. 43, Springer, 1967.

[Q-2] ________, Some Remarks on Étale Homotopy Theory and a Conjecture of Adams, Topology 7 (1968), 111-116.

[Roos] J. Roos, Sur Les Foncteurs de $\underleftarrow{\text{Lim}}$. Applications, C. R. Acad. Sci. Paris 252 (1961), 3702-3704.

[Seg] G. Segal, Classifying spaces and spectral sequences, Publ. I.H.E.S. 34 (1968), 105-112.

[Sieb] L. Siebenmann, Infinite Simple Homotopy Types, Indag. Math. 32, (1970), 479-495.

[Sky] E. Skylyarenko, Homology Theory and the Exactness Axiom, Russian Math. Surveys 24 (1969), 91-142.

[Spa] E. Spanier, Algebraic Tpology, McGraw-Hill Book Co., New York, 1966.

[S -W] ________ and J. H. C. Whitehead, The theory of carriers and S - theory, in Algebraic geometry and topology, a symposium in honor of S. Lefschetz, Princeton U. Press, Princeton, N. J., 1957, 330-360.

[St - 1] N. E. Steenrod, Regular Cycles on Compact Metric Spaces, Ann. of Math. 41 (1940), p. 833-851.

[St - 2] ____________, The Topology of Fibre Bundles, Princeton Univ. Press, 1951.

[St - 3] ____________, A convenient Category of Topological Spaces, Mich. Math. J. 14 (1967), 133-152.

[S - E] ____________ and D. B. A. Epstein, Cohomology Operations, Ann. of Math. Studies 50, Princeton, 1962.

[Sto] R. E. Stong, Notes on cobordism theory, Princeton Univ. Press, Princeton, N. J. 1968.

[Str] A. Strøm, The Homotopy Category is a Homotopy Category, Arch. Math. 23 (1973), 435-441.

[Sul - 1] D. Sullivan, Geometric Topology, Part I, Localization, Periodicity and Galois Symmetry, Mimeographed Notes, M.I.T., (1970).

[Sul - 2] __________, Genetics of Homotopy Theory and the Adams Conjecture, Annals of Math., (to appear).

[Tay - 1] J. Taylor, A Counter-Example in Shape Theory, Bull. Amer. Math. Soc. 81 (1975), 629-632.

[Tier] M. Tierney, Categorical constructions in stable homotopy theory, Lecture Notes in Math. 87, Springer, Berlin-Heidelberg - New York, 1969.

[Tuc - 1] T. Tucker, Some Non-Compact 3-Manifold Examples Giving Wild Translations of R^3, (to appear).

[Ver] J. Verdier, Equivalence Essentielle des Systemes Projectifs, C. R. Acad. Sci. Paris 261 (1965), 4950-4953.

[Vogt - 1] R. Vogt, Homotopy Limits and Colimits, Math. Z. 134 (1973), 11-52.

[Vogt - 2] ______, Boardman's stable homotopy category, Lecture Notes in Math., No. 21, Aarhus Universitet, 1971.

[Wag] J. B. Wagoner, (to appear).

[Wall] C. Wall, Finiteness Conditions for CW- Complexes, Ann. of Math. 81 (1965), 55-69.

[West - 1] J. West, Semi-Free Group Actions on the Hilbert Cube with Unique Fixed Points, (mimeographed notes).

[West - 2] ______, Infinite Products which are Hilbert Cubes, Trans. Amer. Math. Soc. 150 (1970), 1-25.

[West - 3] ______, Mapping Cylinders of Hilbert Cube Factors II - The Relative Case, General Topology and its Applications, (to appear).

[West - 4] ______, Compact ANR's have Finite Type, Bull. Amer. Math. Soc. 81 (1975), 163-165.

[Wh - 1 G. Whitehead, Recent Advances in Homotopy Theory, American Mathematical Society, Providence, Rhode Island, 1970.

[Wh - 2] ____________, Generalized homology theories, Trans. Amer. Math. Soc. 102 (1962), 227-283.

[Yeh] Z. Yeh, Thesis, Princeton, 1959.

Index